# Ocean Energy

Roger H. Charlier · Charles W. Finkl

# Ocean Energy

Tide and Tidal Power

Springer

Dr. Roger. H. Charlier
av. du Congo 2
1050 Bruxelles
Belgium
rhcharlier@hotmail.com

Dr. Charles W. Finkl
Coastal Planning & Engineering, Inc.
2481 NW. Boca Raton Blvd.
Boca Raron FL 33431
USA
cfinkl@coastalplanning.net

ISBN: 978-3-540-77931-5          e-ISBN: 978-3-540-77932-2

DOI 10.1007/978-3-540-77932-2

Library of Congress Control Number: 2008929624

© Springer-Verlag Berlin Heidelberg 2009

This work is subject to copyright. All rights are reserved, whether the whole or part of the material is concerned, specifically the rights of translation, reprinting, reuse of illustrations, recitation, broadcasting, reproduction on microfilm or in any other way, and storage in data banks. Duplication of this publication or parts thereof is permitted only under the provisions of the German Copyright Law of September 9, 1965, in its current version, and permission for use must always be obtained from Springer. Violations are liable to prosecution under the German Copyright Law.

The use of general descriptive names, registered names, trademarks, etc. in this publication does not imply, even in the absence of a specific statement, that such names are exempt from the relevant protective laws and regulations and therefore free for general use.

*Cover design:* deblik, Berlin

Printed on acid-free paper

9 8 7 6 5 4 3 2 1

springer.com

# Contents

**List of Illustrations** .................................................. ix

**1 Poseidon to the Rescue: Mining the Sea for Energy—A Sustainable Extraction** ........................................................... 1
   1.1  Energy From The Ocean ........................................... 1
        1.1.1  Tidal Power ................................................ 2
        1.1.2  Marine Winds ............................................. 4
        1.1.3  Wave Power ............................................... 8
        1.1.4  Ocean Thermal Energy Conversion ..................... 13
        1.1.5  Marine Biomass Conversion ............................. 14
        1.1.6  Marine Currents ........................................... 14
        1.1.7  Tidal Currents ............................................. 15
        1.1.8  Salinity Gradients ........................................ 16
   1.2  Hydrogen Power ................................................. 17
   1.3  Conclusion ........................................................ 17
        1.3.1  Environment Objections ................................. 18
        1.3.2  Bacteria ..................................................... 18

**2 Medieval Engineering that Lasted** ................................. 29
   2.1  Introduction ...................................................... 29
   2.2  Tide Mills, Economics, Industry and Development ............. 30
   2.3  Historical Development ......................................... 31
        2.3.1  The Middle Ages ......................................... 31
        2.3.2  From 1492 to the End of the 18th Century .............. 32
   2.4  Location of Tide Mills ........................................... 33
        2.4.1  Spain ........................................................ 34
        2.4.2  France ...................................................... 35
        2.4.3  Portugal .................................................... 36
        2.4.4  British Isles ................................................ 36
        2.4.5  Northern Europe .......................................... 37
        2.4.6  The Far East ............................................... 37
        2.4.7  The Americas ............................................. 37

|  |  |  |  |
|---|---|---|---|
| | 2.5 | Distribution Factors | 38 |
| | 2.6 | Mills and Their Environment | 39 |
| | | 2.6.1 Dikes | 40 |
| | | 2.6.2 Entrance Sluice Gates | 40 |
| | | 2.6.3 Ponds | 41 |
| | | 2.6.4 Exit Gates | 42 |
| | | 2.6.5 Wheels | 42 |
| | | 2.6.6 The Mill | 43 |
| | 2.7 | Renaissance of the Tide Mill? | 45 |
| | 2.8 | A Preservation-Worthy Heritage | 46 |
| | 2.9 | Conclusion | 47 |
| **3** | **The Riddle of the Tides** | | **65** |
| | 3.1 | What is a Tide? | 65 |
| | 3.2 | Types of Tides | 65 |
| | 3.3 | Tide and Tidal Current | 67 |
| | 3.4 | Tide Generation | 67 |
| | | 3.4.1 Power Generation | 68 |
| | 3.5 | Range and Amplitude | 70 |
| | 3.6 | Transmission and Storage | 71 |
| | 3.7 | Tides and Harmonic Analysis | 72 |
| | | 3.7.1 Smoothing | 73 |
| | | 3.7.2 Auto-Correlation | 73 |
| | | 3.7.3 Moving Averages | 73 |
| | | 3.7.4 Auto-Regression | 74 |
| | | 3.7.5 Fourier Analysis | 75 |
| **4** | **Dreams and Realities** | | **79** |
| | 4.1 | Dreams | 79 |
| | | 4.1.1 The Severn River and Other British Plants | 79 |
| | | 4.1.2 Japan | 81 |
| | | 4.1.3 South Asia, Egypt | 81 |
| | | 4.1.4 Down Under | 81 |
| | | 4.1.5 Much Power, No Users | 82 |
| | | 4.1.6 India | 82 |
| | 4.2 | Realities | 83 |
| | | 4.2.1 The Rance River Plant | 83 |
| | | 4.2.2 The Kislaya Bay Plant (Russia) | 84 |
| | | 4.2.3 Annapolis-Royal Pilot Plant (Canada) | 85 |
| | | 4.2.4 A Hundred Chinese Plants | 85 |
| **5** | **The Anatomy of the Rance River TPP** | | **103** |
| | 5.1 | Introduction | 103 |
| | 5.2 | Ancestors and Forerunners | 104 |
| | 5.3 | Tide Mills Bow Out on the Rance | 105 |

## Contents

|  |  | 5.3.1 | The Rance River Plant | 106 |
|---|---|---|---|---|
|  |  | 5.3.2 | Other Anniversaries | 108 |
|  |  | 5.3.3 | The Anatomy of the Rance River Plant | 109 |
|  | 5.4 | The Rance: First and Last of Its Kind? | | 110 |
|  |  | 5.4.1 | The Past and the Future | 112 |
|  |  | 5.4.2 | Changes at the Rance TPP | 112 |
|  |  | 5.4.3 | Discussion | 113 |
| 6 | **Harnessing the Tides in America** | | | 119 |
|  | 6.1 | The Quoddy and Fundy Affairs | | 119 |
|  | 6.2 | The United States and Tidal Power | | 119 |
|  |  | 6.2.1 | The Passamaquoddy Site | 122 |
|  |  | 6.2.2 | The U.S. and the U.K. | 122 |
|  | 6.3 | Argentina—The San José Tidal Power Plant | | 123 |
| 7 | **Improvements, Adjustments, Developments** | | | 125 |
|  | 7.1 | Taiwan | | 125 |
|  | 7.2 | Gorlov's Barrier | | 125 |
|  | 7.3 | Japan | | 126 |
|  | 7.4 | Russia | | 126 |
|  | 7.5 | China | | 127 |
|  | 7.6 | Great Britain | | 127 |
|  | 7.7 | USA | | 128 |
|  | 7.8 | Norwegian-Dutch Sea and River Mix to Make Power | | 128 |
|  |  | 7.8.1 | Co-Generation | 129 |
|  | 7.9 | Some New Ideas | | 129 |
|  |  | 7.9.1 | Tidal Delay® | 129 |
|  |  | 7.9.2 | Where Do "Things" Stand? | 131 |
|  | 7.10 | Tapping Channel Tides | | 131 |
|  | 7.11 | Turbines | | 133 |
|  | 7.12 | Re-Timing, Self-Timing | | 134 |
|  | 7.13 | Climate Alteration and Energy Shortage | | 134 |
|  | 7.14 | Innovations and New Thoughts | | 135 |
|  | 7.15 | Public Acceptance | | 137 |
|  | 7.16 | New Technologies | | 137 |
|  | 7.17 | Wrap-up | | 138 |
|  |  | 7.17.1 | Does the CEO Get a Pass? | 138 |
| 8 | **Current from Tidal Current** | | | 141 |
|  | 8.1 | Introduction | | 141 |
|  | 8.2 | Tidal Current | | 142 |
|  | 8.3 | Energy Potential | | 143 |
|  |  | 8.3.1 | Regional Potential | 144 |
|  | 8.4 | Geographical Distribution of Promising Sites | | 144 |
|  | 8.5 | Proposed Schemes | | 145 |

|  | 8.6 | Glance at the Past and Look into the Future ..................... 145 |
|---|---|---|
|  |  | 8.6.1 The Modest Forerunners ............................. 146 |
|  |  | 8.6.2 The Contemporary Scene ............................ 146 |
|  | 8.7 | Current Developments ....................................... 149 |
|  |  | 8.7.1 Seaflow and Optcurrent ............................. 149 |
|  |  | 8.7.2 Stingray ........................................... 150 |
|  |  | 8.7.3 Vlieland and the Electricité de France .................. 151 |
|  |  | 8.7.4 In the Arctic ...................................... 151 |

**9 Environment and Economics** ................................... 153
   9.1 Tidal Power and the Environment ............................. 153
   9.2 Economics .................................................. 157

**Annexes** .......................................................... 161

**Annex I: General Bibliography** ...................................... 163
      What was said before 1982 ..................................... 163
      1982–1992 .................................................... 181
      What is Being Said: 1992–2007 ................................. 185

**Annex II: Additional References** ..................................... 193

**Annex III: Special References for Chapter 2** ........................... 205

**Annex IV: Update 2008** .............................................. 207
      Chapter 1 ..................................................... 207
      Chapter 4 ..................................................... 213
      What is a tide? ................................................ 215
      Chapter 9: New Developments ................................. 218
      Chapter 8 ..................................................... 222
      Chapter 9 ..................................................... 223

**Annex V: Companies and Organizations Involved in Tidal Power Projects, Services, and/or Research** ................... 225
      V.1 Equipment ............................................... 225
      V.2 Services, Consultancies and Organizations. ................. 225
      V.3 Various Services and Products ............................ 226

**Annex VI: Summaries** ............................................... 227

**Mini-Glossary** ..................................................... 255

**Index** ............................................................. 257

# List of Illustrations

| | |
|---|---|
| 1 | Tide mill of Bréhat, predilection spot of jules verne and Erik Irsebba |
| 2 | Aerial view of the Rance viwer TPP |
| 1.1 | Schematic of horizontal and vertical axis tidal power turbines |
| 1.2 | Artist's view of turbines in traditional tidal power centrals |
| 1.3 | (**a**) OTEC platform; (**b**) open-cycle OTEC plant (1930) |
| 1.4 | Schematic of open and closed cycles OTEC systems |
| 1.5 | Various pre $21^{st}$-century systems to harness wave energy |
| 1.6 | Cross-section of a typical rim-type generator. (Miller, "Die Straflo Turbine, die technische Realisation von Harza's Idêěn." Zurich: Straflo Group, 1975) |
| 1.7 | Schematic of alternative energy sources |
| 1.8 | Lockheed OTEC scheme. In mid-center: control room; tiny human figures provide dimensions |
| 2.1 | Density distribution of relicts of tide mills in Western Europe. (L. Ménanteau) |
| 2.2 | East Medina mill, Wippingham, Isle of Wight, (Rex Wailes, *Tide Mills in England and Wales*, 1940) |
| 2.3 | Carew, Pembrokeshire, tide mill on Carew River. (Rex Wailes, *Tide Mills in England and Wales*, 1940) |
| 2.4 | Pembroke, Pembrokeshire, tide mill on Pembroke River. (Rex Wailes, *Tide Mills in England and Wales*, 1940) |
| 2.5 | Sluice gate of Birdham tide mill. (Sussex, U.K.) (Rex Wailes, *Tide Mills in England and Wales*, 1940) |
| 2.6 | St. Osyth tide mill, Essex. Stones, wheat cleaner, and sack hoist |
| 2.7 | San José tide mill. 1823. Bay of Cadiz (Map French Military Archives) |
| 2.8 | (**a**) Bay of Cadiz (**b**) St. Banes tide mill, 1823 |
| 2.9 | Tide mill machinery as pictured on a 1703 engraving |
| 2.10 | (**a**) Arillo tide mill on San Fernando-Cadiz road, 1823 |
| 2.10 | (**b**) Arillo tide mill located on road to Cadiz, new facing the sea. (figs. 2.10 belong to Archives of French Land-forces, now in Vincennes, France) (**c**) Present condition Arillo tide mill. (Photo L. Ménanteau) |

| | List of Illustrations |
|---|---|
| 2.11 | Operation of a medieval tide mill |
| 2.12 | Eling mill near Southampton (Engl.) Operating reconstructed tide mill (bakery and museum). (Drawing by Mel Wright) |
| 2.13 | Eling mill, restored 1980 (Ph. D. Plunkett) |
| 2.14 | Traou-Meurmill, Côtes d'Armour (Ph. L. Ménanteau) |
| 2.15 | Grand Traouiéros mill and dike, Tregastet, Côtes d'Armour (Ph. L. Ménanteau) |
| 2.16 | Uregna tide mill (20$^{th}$-century) on Zaporito mole (arch. post-card) |
| 2.17 | Arillo tide mill, Cadiz (Ph. L. Ménanteau) |
| 2.18 | Bartivas tide mill near Chicanadela Frontera. (Photo L. Ménanteau) |
| 2.19 | Bartivas (Ph. L. Ménanteau) |
| 2.20 | Mériadec mill, Badens (Morbihan). (Photo L. Ménanteau) |
| 2.21 | Ancillo mill, Santoñary, Cantabria (Photo Azurmendi) |
| 2.22 | Keroilio mill, Plougoumelen, Morbihan (Ph. L. Ménanteau) |
| 2.23 | Petit Traouiéros mill, Perros-Guirec, Côtes d'Armor (Ph. L. Ménanteau) |
| 2.24 | 17$^{th}$-century Pen Castel mill, Arzon, Morbihan (Ph. L. Ménanteau) |
| 2.25 | Tide mill on the Venera Ria |
| 2.26 | Location map of tide mills in Western Europe |
| 3.1 | Alternative operational modes at La Rance, France |
| 3.2(a) | The Rance River TPP, aerial view |
| 3.2(b) | View of barrage, lock, roadway |
| 4.1 | Location of plants in operation or dismantled, or aborted and sites studied in-depth |
| 4.2 | Major tidal power plant sites |
| 4.3 | Work proceeded at Rance River site inside cofferdams |
| 4.4(a) | View of Rance R. TPP. Chalibert Island is in foreground |
| 4.4(b) | View of Rance R. TPP |
| 4.5 | Location map Severn R. estuary and site proposed TPP |
| 4.6 | Detailed maps of proposed Severn R. TPPs |
| 4.7 | Proposed TPP scheme for the Severn River (Wales) |
| 4.8 | Mock-up of the Kislaya Guba TPP (near Murmansk, Russia) as exhibited in Tokyo by USSR embassy |
| 4.9 | View of the Kislaya Bay TPP |
| 4.10 | Kislaya Bay, USSR. Artist's view |
| 4.11 | Map USSR Tidal Power Sites. Mezen, Kislaya location |
| 4.12 | Powehouse being towed to site |
| 4.13 | Location and artist's view of Kislaya TPP |
| 4.14 | Bult turbine installed at Kislaya TPP |
| 4.15 | Sites of possible Large Russian TPPs and areas of large electrical consumption |
| 4.16(a) | China—Location map of tidal power on Leqing Bay in Zhejiang Province |

List of Illustrations xi

| | |
|---|---|
| 4.16(b) | The Raishakou tidal power station (P.R. China) |
| 4.17 | Location map of Passamaquoddy showind basins of the proposed US-Canada TPP |
| 4.18 | Bulb unit generating caisson |
| 4.19 | Straflo generating caisson |
| 4.20 | Sluiceway caisson |
| 4.21 | Sites map. Australia. Kimberley Region TPP project |
| 4.22 | Korea: tidal power plants location map |
| 5.1 | U.S. President F.D. Roosevelt visits Passmaquoddy Tidal Power Plant Project display |
| 5.2 | The bulb turbine which was installed in tidal power plants in France and Russia |
| 5.3 | Inside the Range River plant, the bulb turbine system is reversible |
| 5.4 | Bulb turbines for high capacity low head power stations |
| 5.5 | *STRAFLO*® turbines for low head and tidal power stations |
| 5.6 | Rance River plant power station: interior view |
| 5.7 | Cross-section of the Rance River plant power station (St-Malo Fr.) |
| 5.8 | Cross-section of the Kislaya scheme (Kislaya Bay Rus) |
| 9.1 | Environmental assessment and impact of tidal power projects |

**Fig. 1.** Tide mill of Brehat, predilection spot of jules verne and Erik Irsebba

**Fig. 2.** Aerial view of the Rance viwer TPP

# Chapter 1
# Poseidon to the Rescue: Mining the Sea for Energy—A Sustainable Extraction

## 1.1 Energy From The Ocean

The first sources of ocean energy that come to mind are the hydrocarbons. From timid extraction operations hugging the coastline and shallow depth wells, not too difficult to cap, giant steps have been made, to the point that platforms have been erected, far out at sea, and oil is obtained from ever-greater depths. The value of methane has become more apparent during the last half-century and gaso-ducts—gas-pipelines—cross ever longer water and land expanses, just as oleoducts, the oil carrying pipelines, do. However, with the urgent need to reduce greenhouse gas emissions, the love affair with gas and oil has considerably tapered down.

The ocean bottom has also yielded coal from mines accessible from land or at sea: Scotland, Taiwan and Japan, for instance, continued ocean coal mining operations. But coal too is not any longer being courted, for the same polluting and global warming causing reasons. Futuristic thoughts go to sophisticated extraction of hydrogen, deuterium, tritium. While these can technically be retrieved, costs are high, prohibitive for many, and technological refinement is still needed. The same is true about the non-renewable sub-marine geothermal energy.

But there are other sources of energy which can and should be put to work, which are non-polluting, and minimally environment impacting. Unfortunately their extraction is often expensive.[1,2] Of these some have been tapped, with unequal success though, such as the tides, the waves, the marine winds, others remain more engineers' dreams like marine currents, salinity differential. As for OTEC, ocean thermal differences, it is technically possible to put it to work, but economically it remains unattractive. To use a French expression, let us have a *tour d'horizon* of the fields.

---

[1] Hislop, D. (ed.), 1992, *Energy Options. An introduction to small-scale energy technologies:* Rugby, Intermediate Technology Publications.

[2] Kristoferson, L.A. and Bokalders, V., 1991, *Renewable energy technologies. Their applications in developing countries:* Rugby, Intermediate Technology Publications.

## 1.1.1 Tidal Power

Anyone who has ever watched tides roll in on the coasts of Normandy or Brittany, on the estuary of the Severn River or in the Bay of Fundy, cannot help but be awed by the force that is unleashed. The phenomenon had, of course, already been observed in Classical Times and this power was put to work on rivers such as the Tiber River in Rome, the joint estuary of the Tigris and Euphrates rivers even much earlier. Tide mills on the Danube may date from later periods. Mechanical power was sought to grind grain, to power sawmills, to lift heavy loads.[3]

These tide mills are of course not different from run-of-the-river mills, except that they include an impounding basin where the water brought in by the incoming (flood) tide is stored: At ebb tide the water is released but has to pass through a channel wherein the mill wheel is set. Some more sophisticated mills even captured power from both ebb and flood tides. And still others captured the energy of the horizontal movement of tides. The tide mills' demise in man's industrial arsenal was slow but their numbers declined rapidly and abruptly, as newer technology unfolded.

The tide mill may appropriately be considered the forerunner of the tidal power plant that generates electricity and, in France for instance, has brought a sleepy region into the twentieth century. The Rance River plant (Brittany) has successfully provided power for more than forty years.[4,5] It has also provided the dismal Russian North with the electricity needed to develop a rather desolate region.[6] The Canadian plant, in Nova Scotia, is more a trial run than a badly needed plant.[7,8] Originally geographically limited to coasts with large tidal ranges, the development of very small head turbines permits the implantation of tidal power plants[9] in many more locations. The development of the tidal power plant went hand-in-hand with, or at least was boosted by that of the bulb turbine (France, Russia)[10] and later of the Straflo turbine (Canada).[11]

---

[3] Charlier, R.H. and Ménanteau, L., 1999, The saga of the tide mill: *Renew. Sustain. En. Rev.*

[4] Charlier, R.H., 1982, *Tidal energy:* New York, Van Nostrand-Rheinhold.

[5] Barreau, M., 1997, 30th anniversary of the Rance tidal power station: *La Houille Blanche-Rev. Int. de l'Eau* **52**, 3, 13.

[6] Bernshtein, L.B. and Usachev, I.N., 1957, Utilization of tidal power in Russia in overcoming the global and ecological crisis: *La Houille Blanche-Rev. Int. de l'Eau* **52**, 3, 96–102.

[7] Anonymous, 1982, *Fundy tidal power update '82*: Halifax, Nova Scotia, Tidal Power Corporation.

[8] Delory, M.P., 1986, The Annapolis tidal generating station: *Int. Symp. Wave, Tidal, OTEC and Small Scale Energy* III, 125–132.

[9] Henceforth referred to by the acronym TPP.

[10] Charlier, R.H., 1982, *op.cit.* fn. 4.

[11] Charlier, R.H. and Justus, J.R., 1993, *Ocean energies*: Amsterdam-New York, Elsevier pp. 316–320.

## 1.1 Energy From The Ocean

The first major hydroelectric plant to use the energy of the tides was put into operation in 1967. It produces approximately 540,000 kW of electrical power[12]. A modest amount in view of heralded plans to produce over a million kilowatts. The dam crosses the estuary of the Rance River at its narrowest point and accommodates a four-lane highway. Bulb turbines permit reversible operation and pumping. The flow of the waters amounts to some $24,000 \text{ m}^3$/sec. The station was linked to France's national electricity grid; this allows to raise the reservoir's level by pumping, thus at high tide the reservoir is overfilled by taking power out of the system, and, at minimal power loss, the reservoir's level is raised 1 m.

The high capital investment required has certainly acted as a principal deterrent to the construction of more tidal plants, and has laid to rest plans for mammoth schemes for the Severn River (Great Britain),[13,14] Chausey Islands (France) and Passamaquoddy Bay–Bay of Fundy. The Chinese government, taking a more down-to-earth view, has constructed over a hundred small plants, using earthen dams, some of which were pre-existing.[15] Government figures disclosed in 1999 at a Qingdao (PRC) conference on the history of oceanography, announced that China's electricity production from tidal energy would reach 50 MW by 2000 and climb to 310 MW by 2010.[16] This would permit electrification of large, but distant, areas. Initial costs were further brought down when it proved possible to construct plants using modules and dispensing with the costly cofferdams. Argentina and Australia, who conducted major feasibility studies, have now been silent for more than ten years on the topic. On the other hand Korea (ROK) had announced serious plans for a large tidal power station, e.g. Garolim Bay, Inchon Bay, and fostered economic studies.[17] A contract with Sogreah, the French company active in hydrological constructions, was canceled for political motives: France's ill-timed diplomatic recognition of North Korea.

It seems that tidal energy could be put to work for poorer nations and regions by using [modernized versions of] tidal mills and modest plants, as did the Chinese.[18,19,20] Furthermore, end of last century studies found that the cost of a

---

[12] The labeling of this plant as "first" requires some caution, as small facilities were installed elsewhere before. The matter is discussed in a later chapter.

[13] Shaw, T.H. (ed.), 1979, *Environmental effects study of a Severn Estuary tidal power station*: Strathclyde UK, The University.

[14] Severn Barrage Committee, 1981, *Tidal power from the Severn Estuary*: London, H.M. Stationary Office.

[15] Cf. fn. 15.

[16] The authors have not been able to ascertain whether this figure was indeed reached by that date.

[17] Chang, Y.T., 1996, Korean experiences in estimating the non-market benefits of the development of coastal resources: the case of a tidal plant: *Book of Extended Abstr. Ocean Canada '96 (Rimouski, Quebec)* 40–44.

[18] Charlier, R.H., 2001, Ocean alternative energy. The view from China—"small is beautiful": *Renew. Sustain. En. Rev.* **5,** 3, 403–409.

[19] Fay, J.A. and Smachlo, M.A., 1982, *Small scale tidal power plants:* Cambridge, MA, Massachusetts Institute of Technology (MIT Sea-Grant College Program).

[20] Cave, P.R. and Evans, E.M., 1984, Tidal energy systems for isolated communities. In: West, E. (ed.), *Alternative energy systems:* New York, Pergamon pp. 9–14.

tidal plant kilowatt is today hardly higher—if indeed it is—than that produced by a conventional central or even a nuclear plant.[21] The longevity of a tidal power plant is between two and three times longer than the lifespan of those.[22]

More modestly even, reintroduction of tide mills in appropriate and selected sites may prove to be a profitable very low cost investment. As proof one may cite several such mills that have been restored and are working musea, or simply an artisanal revival.

Man-made currents can interact with tidal currents to deflect, redirect, modify sediment transport. To the chagrin of dredgers, this would reduce the costs of navigation channel maintenance and control formation of sandbanks hampering ship traffic.

## 1.1.2 Marine Winds

Of all the ocean energies, marine winds have known the most important development during the last decades. They are a "renewable" which was easy to harness and which required only relatively modest capital investments. Sites are abundant, and a judicious choice permits to dampen the objections voiced because of the noise they cause. Marine wind "farms" have been implanted in numerous locations particularly in Northern and Western Europe. However environmental-linked objections are being raised, spurring engineers to devise new approaches.

Most of the ocean energies require engineering developments to be harnessed and produce electricity, except the marine winds and the tides. The WECS, as they were designated a quarter of a century ago, made first a timid appearance, but they were spurred on by ever climbing prices of fossil fuels and the need to reduce carbon dioxide emanations. The technical problems were rather rapidly solved and the first energy captors were erected on land, mountaintops, away from human habitat. The towering structures were not free from environmental impact, particularly noise and aesthetics.

Pylones have become taller and turbines larger. Everyone applauded the harnessing of marine winds but nobody wanted the pylones in his "backyard". There is also concern that the machines may cause hecatombs of birds particularly during migration seasons An answer to noise, migrating birds routes, aesthetics has perhaps been found, at least *in partim:* siting of the marine wind turbines on floating—and movable—platforms. The design is ready and the construction on the books.

It did not stem the determination of some countries to replace by wind, centrals burning coal, oil, or nuclear products. Locations on the coast were favored and even better, offshore sites. From installations involving a few wind turbines,

---

[21] Gorlov, A.M., 1979, Some new conceptions in the approach to harnessing tidal energy: *Proc. Miami Int. Conf. Altern. En. Sources* II, 1711–1795; Gorlov, A.M., 1982, idem: *Proc. Conf. Tidal Power (New Bedford, NS, Inst. Oceanog.).*

[22] Charlier, R.H., 2003, Sustainable co-generation from the tides: *Renew. Sustain. En. Rev.* 7, 187–213; 215–247.

## 1.1 Energy From The Ocean

builders passed to sites where large numbers of turbines were installed. Utgrunden in the Baltic was inaugurated as one of the first "wind farms". The proliferation of marine wind turbines occurred especially in Northern and Western Europe: Sweden, Denmark, Germany, Scotland, The Netherlands, to name a few countries. The success so impressed the Americans that they talked about placing turbines on Georges Bank off the coast of Maine, but it looks like that it is off the coast of Texas that a wind farm will be implanted.

Aeolian energy has been on the foreground for quite some time. The windmill of yesteryear is the undisputed ancestor of today's aero-generator. Wind turbines can of course be installed inland, near-shore or even at sea. Twenty years ago proponents of wind power were derided as a new breed of Don Quichottes.[23] Today even combinations of wind energy parks with coastal defense are being considered. Some thought is being given on capturing offshore winds energy through wind turbines placed along an artificial reef implanted as a recreational beach protection device against waves.[24]

The high population concentration in European countries, their trend to move towards the coasts and the ensuing conurbation restrict the available area. Yet, various studies established that offshore wind resources are far higher than those on land. As water depth increases only slowly with distance from shore along many European coasts this favors mounting of offshore turbines.[25]

Thirteen countries participated in the 2-year assessment project "Concerted Action on Offshore Wind Energy in Europe" (CA-OWEE)[26]; at its issue the view was held that by 2011 the wind parks installed in the coastal seas of Europe[27] might be able to furnish the energy needed by the Union.[28] Some interest has been also voiced in the United States East Coast regions and in Tasmania. All aspects of the problem were considered, including grid integration, but particular focusing was on economics. On-shore-placed turbines are definitely less expensive, so only multi-megawatts centrals would be cost-effective.[29] Higher initial expenses are due to foundations, but also for maintenance and operation.[30] The lion's share of costing is

---

[23] Heronemus, W.E., 1972, Pollution free energy from off-shore winds: *8th Ann. Conf. Expo. Mar. Tech. Soc. (Washington).*

[24] For further information contacts can be made through (1) owner-coastal.list@udel.edu; (2) www.esru.strath.ac.uk/projects/E and E98-9/offshore/wind/wintr.htm; (3) www.coastal.udel.edu/coastal/coastal.list.html

[25] Garrad, M.H., 1994, *Study of offshore wind energy in the EC. Co-funded by the CEC, Joule I Programme:* Brekendorf, Germany, Natürliche EnergieVerlag.

[26] Anonymous, 2001, *Offshore wind energy: ready to power a sustainable Europe*: Brussels, CA-OWEE, The European Commission (Final Report).

[27] Belgian, British, Danish, Dutch, German, Irish, Swedish, possibly French, waters.

[28] Belgium, Denmark, Finland, France, Germany, Great Britain, Greece, Italy, Ireland, Netherlands, Poland., Spain, Sweden.

[29] Cockerill, T.T., Harrison, R., Kuhn, M., et al., 1998, *Opti-OWECS final report.* III: *Comparison of off-shore wind energy at European sites:* Delft NL, Instituut voor Windtechnologie, Technische Universiteit Delft.

[30] Van Brussel, G. and Schöntag, C., 1998, Operation and maintenance aspects of successful large offshore windfarms:*Proc. Europ. Wind En. Conf. Dublin, Ireland* no pp.nbrs

for the turbine (on-shore 71%, off-shore at least 50%), grid connections (on-shore 7.5 %, off-shore 18%) and foundations (on-shore 5.5%, off-shore about 16%). Land turbines cost considerably less than those used with marine installations. The moral of the story is that to reduce costs, the larger the turbine, the better; with rotor diameters of about 70 m a North Sea sited wind-turbine can produce annually between five and six million kilowatt/hour. There being no neighbors to complain of the noise, windmills at sea can safely turn 10–20 % faster than on land.

A park was built eight years ago on the IJsselmeer, in The Netherlands[31]; a second park was inaugurated in 1996 (Medemblik and Dronten). Denmark built parks in 1991 and 1995, but the most recent is at Middelgrunden and is only two years old, and is the largest producer with 89,000 MWh/year.[32] Sweden's installations date from 1990, 1997–1998, and the newest completed recently. The Utgrunden (marine wind-) park (2000) is Sweden's largest with 38,000 MWh/year. The only British facility in operation is located near Blyth and is a relatively small producer with 12,000 MWh/year. Interestingly the Danish Middelgrunden facility is owned jointly by a 3,000+ members wind-energy cooperative and a local electricity utility.

The development of offshore wind farms may however be slowed as the market is liberalized; the cost of the kilowatt must be reasonable at production time or a project's viability will unavoidably be put in jeopardy. It was thus pointed out that Europe may be left in the odd position of disposing of an environment-friendly and abundant energy resource, supported by public and governments alike, but without the market framework to foster its development.

Nine offshore wind farms are planned: five by Denmark (two in 2002,[33] then one each in 2003, 2004 and 2006), one by France (in 2002 near Brest[34]), a near-shore one by The Netherlands (2003), another by Belgium (2003), and one by Ireland[35] on Arklow Bank. Plans in Belgium include, as marine and fluvial installations an additional farm near Zeebrugge and another one along the Scheldt-Rhine canal, north of Antwerp.

A Danish company's subsidiary—Vestas Mediterranean East—will sell some 47 wind turbines (850 kW) to Sicily for three wind projects to Asja Ambiente Italia; the total installed capacity will reach 40 MW and operations started early in 2007. They are dwarfed by the 52 turbine wind farm of Hadyard Hills (South Ayrshire). Thus far 171 MW of electricity generating wind turbines came on line in 2006, providing current for 80,000 household and over 665 MW were added to normal electricity production.

Danish and Dutch projects would produce a kWh for $0.049–0.067 compared to on-shore prices of $0.027–0.07. Production costs vary of course with the speed of

---

[31] Formerly Zuiderzee, prior to the damming and polderization of a major portion of the water body.

[32] Giebel, G., 2001, *On the benefits of distributed generation of wind energy in Europe:* Copenhagen, Fortschritt Berichtte (VDI). DEA/CADETT, 2000, *Electricity from offshore wind:* Copenhagen, Danish Energy Agency.

[33] Two facilities scheduled for that year.

[34] But not built at this writing.

[35] No date set at this writing.

## 1.1 Energy From The Ocean

the prevailing winds, turbine size, and plant dimensions, while technological refinements allow expecting one kWh to cost between $0.04 and 4.6. The Dutch estimate that on their sector of the continental shelf they could erect sufficient wind turbines to satisfy, by 2030 180% of the country's electricity needs. This figure may have to be scaled down, however, as a study conducted in 1995 on behalf of the European Union; indeed, there are several sites where turbines cannot be placed, for instance because depths are too great or the distance to shore is. The same rather simplistic calculation ventured of the possibility that the British could capture at sea four times their electricity needs, the Irish fourteen and the Danes even seventeen.

At Zeebrugge, Belgium, a small park has been installed on the sea harbor breakwaters. Production amounts to 4.8 MW, a drop in the bucket for a country needing 15,000 MZ. Belgian authorities gave recently the green light for positioning fifty air turbines on an artificial island at 15 km (8.10 nautical miles) off-shore from the city-resort of Knokke-Heist.[36] The contractor is *Seanergy*. The 1,000 MW produced are to provide electricity to 85,000 families. Construction is scheduled to start in 2003 and placing into service in 2004. Notwithstanding reports from similar projects concluding to benign influence on the marine environment, an impact study will be conducted. The installation is deemed to have a life span of 20 years and the contractors are held, by the contract, to remove all wastes. However, claiming aesthetic pollution (view cluttering from shore), the city of Knokke-Heist filed an objection with the Council of State to block the construction, even though, to minimize their visibility, the turbines will be painted gray to match the North Sea waters' local color. Coming from a city that has, for many years, notoriously failed to provide adequate water purification facilities, one may raise a somewhat surprised eyebrow....Granted, the marine wind parks are not exactly attractive, yet they are not really objectionable, the more so that they are visible at best as specks on the horizon.

The endless procedures came finally to a close in 2007 and the wind-park will be built. The delay has had one advantage: technology has progressed and the latest and largest turbines will be installed.

Tunø Knob, on the Kattegat (Denmark) towers 40 m above the million kilograms concrete foundation placed at a depth of 3–5 m. Its wings spread about 15 m. But, on the positive side, the sea-turbines are 150% effective compared to their land-placed cousins.

And objections are raised in The Netherlands also, claiming deterioration of the polders' landscape. Yet, the Dutch researchers, buttressed by loud Greenpeace endorsement, estimated already at the end of 1998 that 10,000 MW would be extracted from the North Sea by 2030, or 40% of the current electricity consumption. The government promised that by 2020 the energy used in the country would be generated by sustainable sources.

Wind power from the ocean has also been considered for providing the energy needed by pumps and desalination plants.

Marine winds are providing energy to 24 turbines in Zeebrugge; they have a total capacity of 5.2 MW. An isolated one in Middelkerke is rated at 660 kW, and

---

[36] Federal Minister of the Environment Magda Algoet's decision of June 25, 2002.

five turbines, placed along the Baldwin (Boudewijn) Canal in Bruges have a total capacity of 3 MW.

The German DEWI (Deutsches Windenergie-Institut) conducted an in-depth study which concluded in 2000 that Belgium, The Netherlands, Denmark, Germany, and Great Britain could cover their entire 923 million MWh needs (1999 estimate) from offshore wind-energy. This would, however, require placing 100,000 2-MW turbines in North Sea sites.

Amongst plans often mentioned for Belgium are a wind-farm of 50 2-MW turbines off-shore Wenduine, upgrading of the Zeebrugge "windmills" and addition of two more, the new total of 26 would bring production up from 5.2 to 13 MW.

Such projects, understandably, distress tourism-conscious resort municipalities such as Knokke-Heist and Wenduine-Klemskerke-De Haan.

## 1.1.3 Wave Power

The number of patents taken out on wave power activated machines is stunning, and they go back well over two hundred years. Probably the first to be taken out was by Girard, father and son in 1799 and proposed to take out mechanical energy using a raft. In the twentieth century buoys and lighthouses used wave-generated electricity. In the USA several attempts were made in California (San Francisco, Capitolo, Pacifica). The power is provided by the onslaught of a breaking wave, which can be captured in a reservoir, accessible by way of a converging ramp, and connected with a return channel at the exit of a low pressure turbine. Power can also be generated by means of devices set directly in motion by the wave itself.[37]

Though diffuse, available power is impressive: there is more power represented in the potential energy of a heaving ship than there is present in the thrust of its engines. Summed up the total available power of ocean wind waves amounts to $2.7 \times 10^{12}$ watts. It is conceivable to use similar waves from land-locked seas or even lakes; power of such waves is $2^1/_2\%$ less than that of seawater waves.

Waves are a concentrated form of wind energy. The very nature of wind waves requires a large number of small devices for its energy extraction. Waves have the distinction of making more energy available as energy is extracted, due to the inefficiency at which energy is transferred from the wind to the sea at highly developed sea states.

Engineers and designers have been repeatedly discouraged in their attempts to capture wave energy because the occasionally unleashed fury of the sea destroys stations. The force is such that a 25-ton block of concrete has been found inland, after a storm, at about 5 km from shore. To protect against installations' destruction, special constructions are needed, both quite expensive and return limiting.

---

[37] Wave-harnessing systems can use flaps and paddles, focuses, heaving bodies, pitching and rolling bodies, pneumatic or cavity resonators, pressure, rotating outriggers, surges, a combination of several of the above.

## 1.1 Energy From The Ocean

Research has been pursued on finding appropriate and affordable approaches e.g. in China.[38]

Wave extraction systems utilize either the vertical rise and fall of successive waves, in order to build-up water- or air-pressure to activate turbines, or take advantage of the to-and-fro, or rolling motions of waves by vanes or cams which rotate turbines; or still use the concentrations of incoming waves in a converging channel allowing the build-up of a head of water, which then makes it possible to operate a turbine.

Conversion of energy devices can provide propulsion, buoy power supply, be offshore or shore-based plants. A physics classification would recognize devices that intervene in wave orbits, utilize the pressure field, are accelerative, use horizontal transport from breaking waves. Some 38 systems have been described that fit into four broad types: surface profile variations of travelling deep water waves, sub-surface pressure variations, sub-surface particle motion, and naturally or artificially induced unidirectional motion of fluid particles in a breaking wave.[39]

Stahl had already in 1982 classified devices based on mechanical concepts: motors operated by the rise and fall of a float, by the waves' to-and-fro motion, by the varying slope of wave surface, and by the impetus of waves rolling up a beach.[40]

Converging wave channels, supplying a basin constituting the forebay for a conventional low head power station provide a high output. Their economic feasibility has been repeatedly put in doubt. Generators designed along the lines of conventional aero-generators have been proposed. Waves are commonly available and could be harnessed in far more sites than tides. Numerous large megapolis and conurbations located near the shore would be potential consumers of wave energy, but so would coast sited industries.

The systems involve thus either a movable body, an oscillating column or a diaphragm. Researchers usually cite as advantages of harnessing wave energy that they are pollution free, widely available, a low cost operation, that additional units provide easily additional power, their siting on unused shore-land, installations can double as protective devices for harbors and coasts, generators are more efficient than those of fossil fuel conventional plants, are a power source that is complementary to others, their output is unaffected by weather or climate, the size of waves can be fairly well predicted, potential coupling of stations to desalination plants, benign impact on environment and ecology.

Wave energy has been harnessed recently in sophisticated plants particularly in Sweden and Norway. A comprehensive British study yielded many proposals, but the matter has been, for all practical aspects, been laid to rest. Japan has a very active research program, on-going for decades, which led to some large scale efforts, e.g. the *"Mighty Whale"*, a floating power device with air turbine conversion to

---

[38] You Yage and Yu Zhi, 1995, Wave loads and protective technology of an on-shore wave power device: *Chinese Oc. Eng.* **9**, 4, 455–464.

[39] Panicker, N.N., 1976, Review of the technology for wave power conversion: *Mar: Techn: Soc: J:* **10**, 3, 1–12.

[40] Stahl, A.W., 1982, The utilization of the power of ocean waves: *Trans. Am. Soc. Mech. Eng.* 13, 428–506.

electricity or compressed air[41], or the earlier *Kaimei*, a barge equipped with compressed-air chambers.[42,43]. Air turbine buoys are utilized in Japan—as in the US and the UK—as are air turbine generators (Osaka).

Like for tidal power, there are modest devices that can put wave energy to work and which, consequently are more affordable. In California, close to a hundred years ago, wave power was used to light a wharf underneath which panels had been suspended. At Royan, close to Bordeaux, France, waves provided electricity to a home using an air turbine driven by water oscillation in a vertical borehole. In Atlantic City, New Jersey, floats attached to a pier were activated by horizontal and vertical motion. A Savonius rotor operating pump was installed in Monaco's Musée Océanographique research laboratories. At Pointe Pescade, Sidi Ferruch a low-head hydro-electric plant supplied electricity from a fore bay with converging channels. In Sweden an auto-bailer bilge pump has been placed into service, the sea-lens concept has been developed in Norway, and hydraulic pumping over pliable strips in concrete troughs have been proposed by a Boston, USA firm. Though these approaches were either uneconomical or too small at the time, this may not be the case today. Efforts towards the design of economic devices are being made.[44]

T.J.T. Whittaker reminded his audience, at Queen's University Belfast, in his lecture on the occasion of the award of the Royal Society's Esso Energy Prize, that, for more than 20 years work on wave power harnessing had been pursued in China, Japan, India, Ireland, the United Kingdom, Denmark, Sweden and Norway.[45] Denmark tested some years ago a wave converter. Whittaker stated that wave power is a potentially viable technology that could make a significant contribution, to not only European, but also the world energy demand.

A somewhat similar *son de cloche* has been heard in the United States. Indeed, the US Electric Power Institute reports that wave power may be economically viable, but would need a production volume of 10–20 GW. Hawaii, Northern California, Oregon and Massachusetts are proposed as the best sites. It even expressed a preference for waves to wind because of lesser visibility and lower profile in addition to better dispatchability. American researchers concluded that to make such significant contribution sustained research is needed into the application of wave power to the offshore production of hydrogen. The State of Oregon set up a National Wave Energy Research, Development and Administration center; it is part of Oregon's effort

---

[41] Hotta, H., Washio, Y., Yokozawa, H. and Miyazaki, T., 1996, Research and development on the wave power device "Mighty Whale": *Ren: En:* **9**, $^1/_4$, 1223–1226.

[42] Kudo, K. and Hotta, H., 1984, Study of the optimal form of Kaimei-type wave power absorbing device: *ECH. Rep. Jap. Mar. Sci. Technol. Center* 13, 63–84.

[43] Cf. Charlier and Justus, 1993, *op. cit.* pp. 136–140

[44] French, M. and Bracewell, R., 1996, The systematic design of economic wave converters. In Chung, J.S., Molagnon, C.H. and Kim, A. (eds), *Proc. 6th Int. Offshore and Polar Engng Conf. (ISOPE CO)* I, 106–110.

[45] The lecture was delivered in 1995. Recent publications on wave power in India include e.g. Raju, V.S. and Ravindram, M., 1996, Wave energy: power and progress in India: *Ren. En.* **9**, 1–4, 339–345, and the assessment of wave power potential for the Indian coasts by Sivaramakrishnan, T.T., 1992, Wave power over the Indian seas during southwest monsoon: *Energy* **17**, 6, 625–627.

## 1.1 Energy From The Ocean

to kindle marine renewable and sustainable energy systems. Thus $5 million will buttress the a*d hoc* programs conducted by Oregon State University.

India's Institute of Technology considered combining a wave energy converter with a fishing harbor breakwater, thereby making double use of the concrete works, as suggested by Whittaker and this author (Charlier). The Indian researchers of IIT also developed a power system using the piezo-electric effect: plastic sheets are to be suspended from floating rafts and secured to the ocean bottom. As waves lift the rafts, the sheets bend and generate electricity in the process.

Among the more recent devices due for deployment *in situ* is the Pelamis P-750 Wave Energy Converter, tested since 1998, was placed on the market by Ocean Power Delivery Ltd®, a Scotland based company. A full-scale pre-production prototype was built in 2003, and field-tested in 2004. The 750, in the model's name, refers to 750 kW power.

Several Pelamis have been installed in a limited make up "wave farms"+ (Fig. 1.1) similar to "wind farms", "biomass farms", "fish farms", "oyster and mussel parks". This first try-out will take place in the Orkney Islands located European Marine Energy Centre.

The number of such machines required to offer a significant saving of traditional fuels, is however rather large, the space required not minimal. The company views a field of 1 to 2 km$^2$ wherein 40 Pelamis would be installed. The total output of the farm, 30 MW, is potentially sufficient power to fill the needs of 20,000 homes (Fig. 1.2).

The Pelamis device belongs to the group of semi-submerged articulated structures, of which other types have been tested and proposed in the past.[46] Pelamis heads on into the incoming wave and contains three 250 kW-rated power conversion modules, each a ≪generator≫ in its own right. Hydraulic arms resist the wave motion which pumps an intermediary fluid through motors by the way of smoothing accumulators. A single dynamic umbilical conduit is connected to the nose-located machine's tranformer leading the power to the seabed.

It is ≪sustainable≫, non-site specific, has good power capture efficiency, deployable in depths up to 100 m, is price competitive with an offshore wind power scheme, and an eventual lower kWh generation price is predicted by its manufacturer. Yet, some of the objections voiced against wave energy conversion schemes, and occasionally confirmed by experience, remained unanswered and proponents would gain support if addressed. First the WEC scheme's vulnerability to exceptional storms, next the obstacle the WEC constitutes to navigation. Except for Scandinavia, wave power had somewhat slid into a forgotten corner of Neptune's power potential. Nothing would prevent, except perhaps the need for space, this

---

[46] Charlier, R.H. & Justus, J.R., 1993, *Ocean energies. Environmental economic and technological aspects of alternative power sources*: Amsterdam, New York, London & Tokyo, Elsevier [Oceanography Series Nr 57] pp. 122–153; Ross, D., 1981, *Energy from the waves: the first ever book on a review in technology*: New York, Pergamon; Salter, S.H., 1979, Recent progress on ducks: *Symposium on wave energy utilization-Chalmers University of Technology, Göteborg, Sweden.*

technology to be adapted to the limnology domain, even if schemes could conceivably have to be more modest in size.

A wave farm has been placed off the Portugal coast in 2006. The Archimedes Wave Swing generator—designed and developed by a Scottish company—completed successful trials in Portguese waters. The system is moored to the seabed and is invisible from the surface. Electricity is generated as waves move an air-filled upper casing against a lower fixed cylinder. The technology is Dutch in origin. The nearly €3 million input allows the completion of a full-scale plant that could be on-line by 2008.

The first M/V Sea Power was installed in late 2006 at a site some 7 km off the coast of northern Portugal, near Póvoa de Varzim. Ocean Power Delivery (OPD) signed a contract with a Portuguese consortium, led by Enersis, to build the initial phase of the world's first commercial wave-farm to generate renewable electricity from ocean waves.

The 2.5 MW project is expected to meet the electricity demand of more than 15,000 Portuguese households while more than 60,000 tonnes per year of carbon dioxide emissions from conventional generating plants will be displaced.[47]

On October 1, 2006 wave powered electricity for 1,500 families in Portugal was provided by a floating electric central sited some eight km offshore from Aguçadoura. Rui Barros, director of Enersis, is reported to have announced the central being placed on line as a world's first. Wave power has been used, however, for close to a century in Royan, Monaco, a pier had been lit by wave energy in Pacifica, California, a beach had hosted a simple machine, systems had provided mechanical power, etc and pilot plants provided current in Scotland and Norway to mention just two locations.[48]

It remains nevertheless gainsaid, that it is the first time wave energy has left the endless academic discourse and timid try-outs area, and be put resolutely to work. The Ocean Power/Enersis system encompasses three 3.5 m diameter 142 m long pipes, three generators and a set of hydraulic high-pressure pumps. Generated current is led to the continent via submarine cable. Refining of meteorological equipment and methods currently allows prediction of force and height of waves up to six to seven days in advance.

Costs are about the same as that of a wind-system—an approach to alternative energy already endorsed by Portugal earlier—but optimistic prognoses of the designers assert that the wave farm will yield three times that of the wind farm. The same optimists plan to establish 28 more floating centrals by mid-2008.

Besides wind and waves, the Portuguese are also eyeing the sun as an alternative source of energy. They started construction, in 2004, of what may well be the largest photovoltaic energy conversion plant, intending to connect no less than 100 hectares of sun-panels.

It is not always a matter of producing domestic or industrial electricity. The up and down movement imparted by waves to a ship produces power than considerably

---

[47] *Cf.:* www.greenjobs.com/Public/IndustryNews/i_news_00411.htm; www.oceanpd.com; www.google.nl/search?hl=nl&q=wave+farm+portugal&btnG=Google+zoeken&meta

[48] Charlier, R.H. and Justus, J.R., 1993, *Ocean Energies*: Amsterdam, Elsevier.

exceeds that power to propel a ship; technology that would allow to convert one into the other would reduce considerably transportation costs. Some ships likewise have added to their upper-structure panels to absorb marine wind power. In Mexico experiments were conducted on a wave-powered pump system to flush stagnating water in foreshore lagoons. Ireland concentrated on oscillating water column systems. Other pumps have been designed by Isaacs of La Jolla (California) Scripps Institution of Oceanography—tested off Kanoehe Bay (Hawaii). The European Union contributed to the funding of an oscillating water column plant to substitute, on Pico (Azores) wave power to diesel. Of all the devices proposed and researched in the United Kingdom, only two were retained for further studies: an oscillating water column (OWC) and the circular "Sea Clam". The OWC was deployed on the Island of Islay utilizing a natural rock gully, thus saving on construction outlay and facilitating maintenance access. Another European Union funded project is a near-shore sea-bottom sited two-chamber OWC in Scotland. The University of Edinburgh was the site of S.H. Salter's "nodding duck" (rotating vane) research. Belgium examined a decade or so ago, the possibility to use wave power to reduce silting in the harbor of Zeebrugge.[49]

In Toftestallen, Norway, the world's largest oscillating water column system had a capacity of between 500 and 1,000 kW. It functioned properly but was unfortunately wrecked in 1998 during a particularly heavy storm. It has not been reconstructed thus far (2007).

## *1.1.4 Ocean Thermal Energy Conversion*

Sometimes referred to as thalassothermal energy [conversion], commonly designated as OTEC. The OTEC uses the difference of temperature prevailing between different ocean waters layers to produce electrical power. Statisticians eager to impress the amount of energy available stress that in the waters between the tropics the quantity of heat stored daily by the surface water layers in a square kilometer equals the burning of 2,700 barrels of oil.

The pilot projects of Arsène d'Arsonval and Georges Claude have been abundantly and repeatedly described; they date back to the first half of the last century.[50] Following the oil crisis of 1973, there was a new flurry of interest for OTEC and "Mini-OTEC" and "OTEC-1" were launched respectively in 1979 and 1980. In 1981, Japanese researchers built a close-circuit central on Nauru that delivered 31.5 kW/h; they had placed the cold-water conduits on the ocean floor at a depth of 580 m. It was a result that went way beyond the most optimistic expectations.

Several technical improvements have been introduced into the plans of proposed schemes. Energy conversion reaches an efficiency of 97%, water exchanges are no longer made of titanium, but of the far less expensive aluminum, corrosion and

---

[49] Charlier, R.H. and Justus, J.R., 1993, op. cit.
[50] id. fn. 30.

bio-fouling have been considerably reduced, and the closed circuit system is far more ecologically benign than the open circuit one. The 1993 closed-circuit prototype set up at Keahole Point (Hawaii) delivered 50 kWh net. Turbine improvements are under scrutiny.

These very small plants, alas, produced electricity at high cost. Newer systems have been developed by TRW and Lockheed, but have not been tried out. TRW's 103-m diameter concrete emerging platform tops four OTEC units connected to a single cold water adduction pipe plunging to a depth of 1 200 m. Ammonium gas is the intermediary fluid that is considered currently. On the other hand no platform is foreseen in the Lockheed scheme that also consists of four units connected to a concrete column reaching a depth of 450 m.

OTEC facilities could be coupled to desalination plants, aquaculture schemes, air conditioning systems. The Hawaii Ocean Science and Technology Park, on the island of Hawaii (the "big island") is the site of deepwater intake pipes for aquaculture operations; it is also the locale of significant alternate energy research where *i.a.* experiments are currently—and have been for some time—conducted for ocean thermal energy conversion. It is furthermore on Hawaii that research progresses on seawater use for air conditioning, for a variety of alternate energies, and where recently a new study group has perfected plankton growth for the manufacture of bio-fuels.[51] It is also on Oahu (Hawaii) that an "EnergyOcean 2007" was held from August 21 through 23, 2007.[52]

While research is proceeding on a modest scale, no full-size OTEC-central has ever been built nor placed into service.

### 1.1.5 Marine Biomass Conversion

Little new has been reported in the area of marine biomass conversion even though the increase in algal biomass has caused serious concern to coastal regions, and in particular to resort towns. This is in opposition to the considerable progress made with biomass utilization for other purposes than electricity production.

Experts hold that the marine biomass conversion holds promise, has a future but predict that its development will be rather on a regional level, and on a modest scale.

### 1.1.6 Marine Currents

There is no arguing that ocean currents represent an enormous energy potential. To harness it, there has been no shortage of proposals. Some projects envision turbines that are fixed on the seabottom, others would place them in the current itself, allowing several turbines to be attached at different depths to a single cable. As dis-

---

[51] See further in Sect. 1.1.5.
[52] info@energyocean.com and www.Ocean-techexpo.com

tances to the consumer might be, in some instances, too great, industrial complexes were proposed in the middle of the ocean and the manufactured product would then be brought by ship to the continent.

A Canadian concern after testing six prototypes decided to construct a 2,200 MW ocean current energy conversion plant in the Philippines using a Davis Hydro Turbine. The scheme foresees a dam wherein a number of slow rotating vertical turbines are to be housed.

The projects clash however with concerns about navigation safety, climate modification, danger for ocean life, cleaning of floats if they were used. After rejecting the idea of harnessing the Mediterranean's waves—their height being far more modest—Italians are again considering a marine current central in the Straits of Messina.

## 1.1.7 Tidal Currents

Should tidal currents be discussed as part of tidal power or as a special type of marine current? The horizontal to and fro current due to the tidal phenomenon may be tapped in rivers as well as in estuaries or bays. It has thus seemed more logical to treat it separately here; tapping tidal current power has received recently more attention even if it has been a provider of mechanical energy in earlier times (tide mills).

Considering tidal currents, rather than tides themselves, poses new problems both from an environmental point of view and of that of power production. Considerations are in order because over the last two or three years there has developed (again) a real interest in tapping such currents for electricity production.

Robert Gordon University (UK) professors Bryden, Grinsted, and Melville have directed susbtantial efforts since the start of the new millennium in making possible a way to extract energy from the tidal current.[53] In a recent paper (intended for the *Journal of Applied Physics*), they developed a simple model to assess the influence that extraction of energy could have upon flow hydraulics. Ten percent extraction of raw energy would result in flow characteristics modifications, and could be used as an approximate guideline for the resource potential of a tidal energy extraction site.

Even though subject to meteorological vagaries, tidal currents, like tides, are an essentially predictable, sustainable and renewable source of energy. If in Scotland Spring tides may provide a kinetic energy flux of $175\,\text{kW}/\text{m}^2$ there are many more regions throughout the world where the flux is about $14\,\text{kW}/\text{m}^2$ which is sufficient for power production. Unlike atmospheric currents, tidal current fluxes are constrained between the seabed and the sea surface, may even be further constrained

---

[53] Ian Bryden, now (2007) at the University of Edinburgh, see following fn., chaired a session dubbed "Wave and Tide Farming" at the conference "Oceans 2007, IEEE/OES, Marine Challenges: Coastline to Deep Sea", held in Aberdeen, Scotland 18–21 June 2007. [IEEE=Institute of Electrical and Electronics Engineers (UK); OES=Oceanic Engineering Society].

in a channel. Hence identification between wind, particularly marine winds, currents and tidal currents is hardly appropriate.

There is a steady decrease in depth and increase in flow speed along a channel, but when energy extraction occurs, a substantial head drop develops where the extraction of energy takes place and flow speed decreases. In the Robert Gordon University model calculations are based upon 10% extraction at 2 km from the channel entrance. Obviously, energy extraction has a negative (reducing) effect on flow speed.

From a practical viewpoint it appears thus not possible to predict energy production only based upon natural river flow. The authors point out that in more complex systems, e.g. the Stingray, two, even three dimensional flow analyses are appropriate.[54]

## 1.1.8 Salinity Gradients

Membrane problems, particularly their cost, remain a major obstacle to progress in tapping that sort of ocean energy. A recent proposal led to the development of a prototype scheme wherein the surface of the ocean plays the role of membrane. In a nearby area fresh water can be stored. Based upon the osmosis principle, it will migrate in the direction of the salty seawater mass, passing through a turbine and mixes with the seawater on the other side. A handicap is the size of turbines required, but if salinity power has to be generated, this seems, today the least expensive approach.

The salinity gradient has been used for electricity production through batteries. The principle involved is reverse electro-dialysis; alternating cells of fresh and salt water are placed next to one another. Flowing seawater take on the role of electrolyte. Lockheed built a 180 MW experimental central. Such batteries are voluminous and the system uses up a good part of the produced current to activate the water pumps.

From an environmental viewpoint the use of salinity gradients does not appear to be free of problems: animals are apt of being sucked-up in the conduits, salt residues must be properly disposed of, and would sufficient fresh water be available in a time when it is at a premium.

Efforts to tap salinity differences may include use of dry holes drilled in the course of the search for oil wells that uncovered brines and brackish water "deposits".

---

[54] Bryden, I.G., Bulle, C., Baine, M. and Paish, O., 1995, Generating electricity from tidal currents in Orkney and Shetland: *Underwater Technology* 21, 2, 17–23; Cave, P.R. and Evans, E.M., 1984, Tidal stream energy systems for isolated communities. In West, M.J. et al., *Alternative energy systems. Electrical integration and utilisation:* Oxford, GB, Pergamon Press; Macleod, A., Barnes, S., Rados, K.G. and Bryden, I.G., 2002, Wakes effects in tidal current turbines. In MAREC, Marine renewable resources *conferences*, Newcastle, September 2002. Bryden, I.W., (in press), Assessing the potential of a simpled tidal channel to deliver useful energy: *J. of Appl. Phys.*

## 1.2 Hydrogen Power

Close to 150 years ago, French fiction writer Jules Verne pictured in one of his books a world in which not coal but water would fuel machines and heat homes. The dream is to come true. In Iceland a battery with "combustible" hydrogen is in the works. The battery has been appropriately labeled a mini chemical factory: Electricity generation is achieved with water vapor as waste product, and hydrogen as fuel. Hydrogen is a main component of water and can be extracted from seawater, for instance, by electrolysis. The extraction, however, requires energy and it is of paramount importance that such energy be "clean" and that the process be sustainable. As mentioned above, American researchers have been thinking of wave energy to power the system.

If a hydrogen battery has been placed on the market[55], it uses natural gas as a driving force and thus pollutes, though less than 50% than a conventional gas-vapor turbine. Japanese workers are placing hopes in photo-catalysts that permit electrolysis by sun-power. Even more unusual is the project that intends to utilize an alga which extracts hydrogen and whose multiplication has been tripled through genetic manipulations developed at Bonn University (Federal Republic of Germany). Probably both once-Bonn resident L. van Beethoven and world-dreamer/science fiction author Jules Verne would be thrilled by the developments.

## 1.3 Conclusion

Several ocean sources of energy can and have been used to produce mechanical and electrical power. If electricity generation is currently the main concern, applications in derived domains, e.g. desalination, or other fields, e.g. pumps, buoys are frequently wave energy activated. The oceans' energies can be put at work in industrialized and developing countries alike. Mega-projects are costly and not necessarily guaranteed of being profitable. On the other hand there exist many possibilities to set up small scale schemes. Other sources have occasionally been tried, but their economic success is still in doubt, while still others appear to hold little promise, at this time, for a reasonable cost implementation. Though a considerable resource, several energies of the ocean are not, any more than oil (petroleum) or gas, inexhaustible. Heat is extractable, thorium and uranium fission, deuterium and hydrogen fusion are potential power or power-related sources. Of the latter hydrogen exceeds the "lifetime of the sun" and deuterium nears it.

Conflicts between aesthetics and lowering of air pollution reduction should, in these authors' view, be resolved in favor of cleaner air and lower air warming. The coastal protection role that artificial islands would play has hardly been mentioned and thus secondary use of the islands apparently not taken into account in calculating the price of a kilowatt.

---

[55] *Cf.* www.promocel.be; www.clubpac.be; www.ulg.ac.be.

## 1.3.1 Environment Objections

Conflicts between aesthetics and lowering of air pollution reduction should, in this author's view, be resolved in favor of cleaner air and lower air warming. The coastal protection role that artificial islands would play has hardly been mentioned and thus secondary use of the islands apparently not taken into account in calculating the price of a kilowatt. There is, however, a general enquiry under way into the building of artificial reefs that would be used as a coastal defense scheme and a site of wind farms.[56] Very little has been said about the coastal defense role that constructions related to ocean energy tapping could play.

## 1.3.2 Bacteria

If the hydrogen of seawater has been repeatedly suggested as intermediate fuel, or as a way of "re-timing" tidal power, hydrogen has been "produced" from other sources. Among such substitute power "sources" a bacteria has been identified that is able to produce hydrogen from sugar-saturated wastes, as proven by a test using wastes from a candy (nougat) and beverages (caramel) factory. Hydrogen and organic acids produced during a first phase, and the acids, are then converted into a hydrogen source by the action of another bacteria. A battery is then the source of electricity. As for the carbon dioxide produced in the first phase' it is removed from the process.

This method was developed at the University of Birmingham (UK). Somewhat akin to it, at the University of Wisconsin (USA), a synthesis process was discovered for an aromatic aldehyde from fructose. Nothing new in fact, as the process is an update of one developed before 1930. It is now cost-wise less onerous because productivity of the catalytic synthesis reaction is speeded up. Direct manufacture of hydroxide-methyl-furfural from agriculture products is based upon separated and purified sugar dehydration. It is achieved in aqueous phase with a catalyst to avoid parasitic reactions.

These developments allow the somewhat tongue-in-cheek remark that ocean energy sources are certainly not "stranger" than bacteria.

---

[56] www.coastal.udel.edu/coastal/coastal list html; owner-coastal list @udel.edu; licofia@att.net

## 1.3 Conclusion

**Table 1.1** Renewable and sustainable coastal zone alternative energy

Energy/power alternatives from ocean sources

In use
- Winds
- Waves
- Tidal currents
- Tides

Potential
- Ocean currents
- Temperatures differentials
- Biomass
- Salinity differentials
- Others

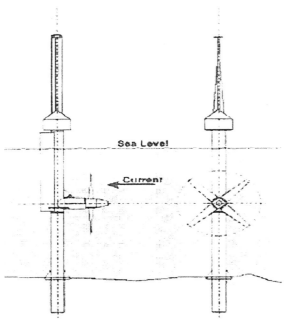

**Fig. 1.1** Schematic of horizontal and vertical axis tidal power turbines
Source: Fujita Research http://www.fujita.com/archive-frr/TidalPower.html

**Fig. 1.2** Artist's view of turbines in traditional tidal power centrals
Source: Energy Authority of NSW Tidal Power Fact Sheet

a. bulb-turbine

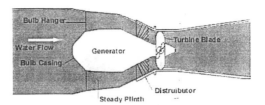

b. Straflo-turbines

   1. Rim-turbine

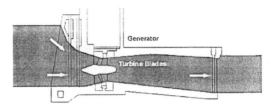

   2. Table-shaped-turbine

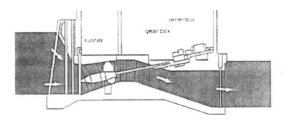

**(a)**

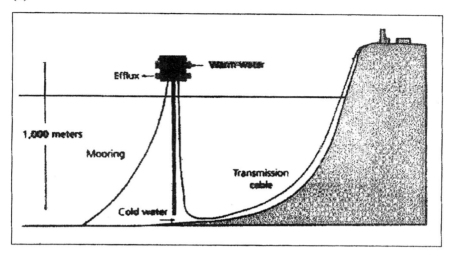

**(b)**

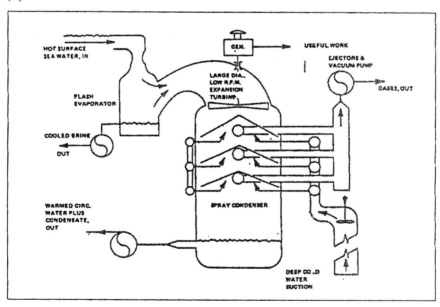

**Fig. 1.3** (a) OTEC platform; (b) open-cycle OTEC plant (1930)
Source: (a) The International Council for Local Environmental Initiatives, http://www.iclei.org/efacts/ocean.htm
(b) NSF/Nasa Solar Energy Panel

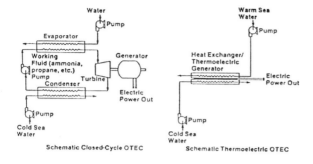

(a) Open-cycle OTEC plants.

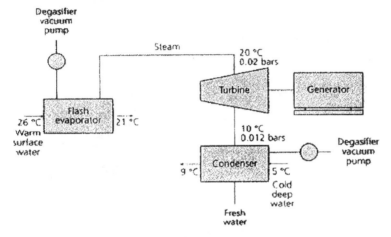

(b) Closed-cycle OTEC plants.

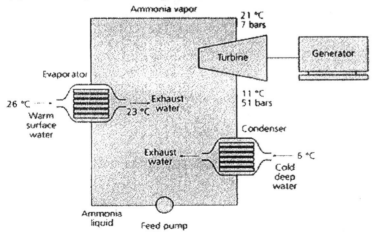

**Fig. 1.4** Schematic of open and closed cycles OTEC systems
Source: The International Council for Local Environmental Initiatives, http://www.iclei.org/efacts/ocean.htm

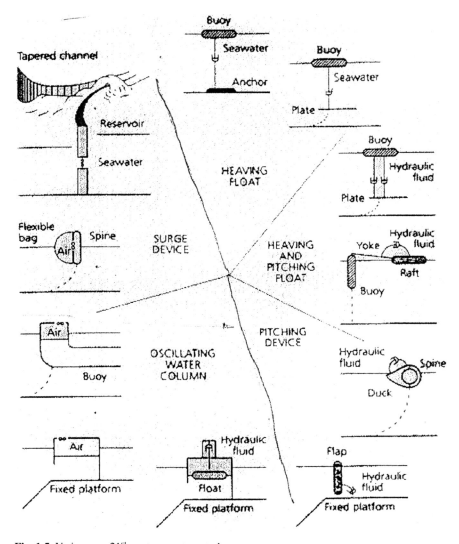

**Fig. 1.5** Various pre 21st-century systems to harness wave energy
Source: The International Council for Local Environmental Initiatives, http://www.iclei.org/efacts/ocean.htm

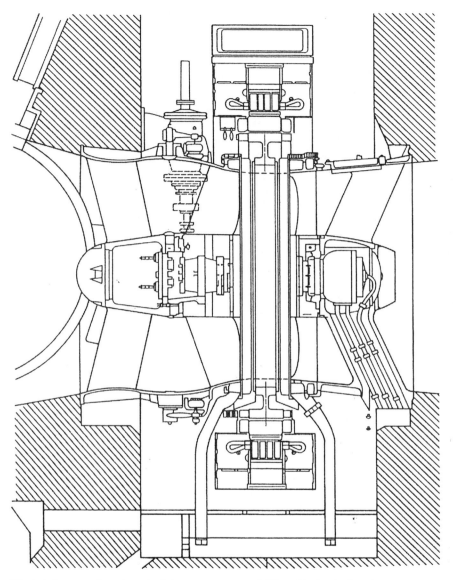

**Fig. 1.6** Cross-section of a typical rim-type generator. (Miller, "Die Straflo Turbine, die technische Realisation von Harza's Idêêñ." Zurich: Straflo Group, 1975)

## Table 1.2 Alternative Energy Systems

| System | Description | Cost/Requirements | Status/Findings | Outlook |
|---|---|---|---|---|
| Solar Photovoltaics | Solar radiation converted into electrical power directly without intermediate step involving heat. | $10-$16 per peak watt of output (over 10 times cost of conventionally generated electricity). | ERDA has developed prototype 1-kWe array using acrylic Fresnel-lens concentrator. Efficiency of new cells are up to 16 percent. Array also produces 5 kW of energy as hot water. | Has significant possibilities for further development and economic breakthroughs. Signs are that inexpensive amorphous silicon hydride films could cut cost of power to 30¢/peak watt. |
| Solar Thermal Water Systems | Water, usually over black surface, under glass, is heated by sun, circulated. | Energy storage. High equipment cost. Solar system to supply half of average home's electrical needs costs $20,000 and water system costs $6,000-$8,000. | Solar thermal energy could supply an estimated 2% of U.S. after 2000. Avg cost of 10¯¹ kW of heat solar power hits earth daily. | OECD says 720 million BOE from solar thermal power by 2000. Heat storage system is necessary to increase effectiveness. |
| Solar Satellites | Massive arrays of solar photovoltaic cells in geosynchronous orbit transmit energy via microwave to receiving station 36,000 mi. below. | More R&D, costs studies needed. High costs; evaluation of environmental effects needed. | NASA says 10,000 - MWe space solar "parks" possible for 2000 and beyond. | None; refer to 2000. |
| Solar Cooling | Absorption-type refrigeration and cooling systems. | Efficiency and economy must be improved. | Most effective in tropic zones where supply/demand are closely matched. | Outlook is good tapering next century. |
| Waves—Relative Motion Power | Floating plants. Wave motions translated into reciprocal motion. Vertical floating plates create liquid pumping action. | Varying wave heights. Expensive to install and corrosive. | Power potential of average wave = 40 mW/km of beach. Salter estimates total UK electric needs could be met with wave energy. | UK scientists believe system will be feasible by 2000. Wave power on UK west coast is 4 times demand for electrical needs. |
| Waves—Air Column Turbines | Column of water rises and falls in piston-like chamber. Runs generator. | Large structure costs $30-$120,000 in steel or concrete. Mooring problems. | Japanese, UK, U.S. designs produce 5 kWe via miniature device. Extraction of 50% wave energy possible in tests. Cost installed: $2,300-$6,740/kW. | Now used for fog horns, etc. |
| Wave-actuated Turbines | Wave spills into core of submerged turbine. Vortex of swirling water acts as flywheel. Most prominent is Lockheed's Dam-Atoll. | Cost of concrete structure, mooring. | One 250-ft diameter Dam-Atoll to produce 1-2 mW on average. All types have damping effects on waves; possible harbor advantages. | May have promise. Prototype engineering underway. |
| Tidal Power | Dam across tidal basin drives generator by water flow. | Geographically limited to areas with large tidal range. | French La Rance facility producing 8,000-1,450,000 kWh/yr since 1961. Other likely areas: Bay of Fundy (Canada), Kislaya Bay (USSR), Severn Estuary (UK) and Scottish lochs. | Substantial in limited areas. Costs are tremely high with current technology. 3 million MWe possible with dams. |
| Ocean Currents | Giant, low-speed turbine powered by Gulf Stream, Kuroshio. Electricity cabled to shore. | Some concern over too much exploitation slowing current. Structure cost. | Power potential of Gulf Stream: 15 gigawatts, Kuroshio—50 gigawatts. Prototype planned in U.S. | Negligible prior to 2000. |
| Grains, Sugar Cane | Fermented into alcohol fuels—ethanol, methanol. Ten percent mixed with gasoline makes "gasohol." | But: Biomass is 50-90% water. Low energy input/output ratio. Use of grain food plants must compete with plant use as food. | Sugar cane yields 4 billion liters/yr. (2.5 million b.o./yr.) in Brazil. | Viewed as significant energy source but doubtful in many areas because of competing land use factor and need as food. |
| Kelp and Seaweed | Giant kelp, other seaweeds, farmed on submerged floats offshore. Nutrients brought from depths. Anaerobically digested to methane gas. | Establishing kelp farms of significant size. Cost of offshore facilities. | 165 to 335 sq. mi. of farm could produce 20 Tc methane gas, according to the Gas Research Institute study. Cost: $3-$6 per million Btu's. | Could supply 100 million BOE after 2000. Kelp farm off southern California now operating tests. |
| Energy Tree Farms, Waste Wood | Special high-yield trees, forest waste can be formed into compressed fuel units. 1 cord wood = 3.5 BOE. | More research needed. Planting large areas; slow growth. US. removes land from agriculture. | Canada, Sweden and U.S. are considering large-scale projects. U.S. State of Georgia alone could grow 50 biomass/yr. (mostly wood). | Very promising—Georgia Tech. |
| Diesel Trees | Diesel fuel can be extracted directly from the South American Copaiba tree, naphtha from Euphorbia tree (sembada). | Requires 10 to 15 years for trees to mature; requires tropical to sub-tropical climates. | Yields: Copaiba—80 liters/yr. Euphorbia—10 bbl/acre now, expected to improve to 100 bbl/acre with development. | Substantial contribution possible after 1985. |

**Fig. 1.7** Schematic of alternative energy sources

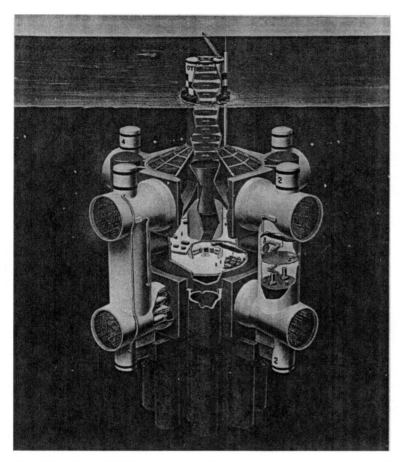

**Fig. 1.8** Lockheed OTEC scheme. In mid-center: control room; tiny human figures provide dimensions

# Chapter 2
# Medieval Engineering that Lasted

## 2.1 Introduction

The use of the energy generated by tides[1] is an integral part of the history of the sea. After France built its sizeable tidal power plant on the La Rance River, near Saint Malo in Brittany, in 1966, the former Soviet Union, China and Canada followed suit with smaller ones, thereafter seemingly a curtain fell on tidal energy harnessing schemes. One would be remiss, however, not to mention Chinese claims that they built a power station before the French La Rance central, that Korea came very close to constructing one (plans coming to naught for international political reasons), and the Bostonians mention a tide-capturing station at the end of the nineteenth century falling victim to port extension.

The recent interest in sources of renewable energy has undoubtedly contributed to the present attention given to the forerunners of tidal power stations.[2] The historic value of tide mills is being recognised, perhaps buttressed by possibilities of reviving their use towards modern versions. Some have indeed been put back into working conditions and the Southampton one is now "a working museum". Changing attitudes towards industrial archaeology[3] and growing concern for our maritime architectural and environmental heritage have also influenced the present trend towards the study and conservation of these remains, proof of the ingenuity of our forebears.

So far, the study of tide mills, on both sides of the Atlantic, has been very uneven. In Britain, Rex Wailes published the first detailed study of the mills of England and Wales in 1941.[4] A few years earlier, a brief paper on the mills of the Basque Country

---

[1] Mariano, *Utilization of tidal power* [in latin], Siena, 1438.

[2] Roger H. Charlier, "From TideMills to Tidal Power", in: *Tidal Energy*. New York London Melbourne, Van Nostrand Rheinhold, 1982, 2: 5274; *idem*, "Chapter VII", in: R.H. Charlier, & J.R. Justus et al., *Ocean resources; Environmental, economic and technological aspects of alternative power sources*. Amsterdam, London, New York, Tokyo, Elsevier, 1993.

[3] Robert Angues Buchanan, *Industrial archaeology in Britain*, London, Penguin books, 1977, 446 p.; Maurice Daumas, *L'archéologie industrielle en France* (Les hommes et l'histoire), Paris, R. Laffont, 1980, pp. 347–396 (Chapitre Les moulins de marée).

[4] Rex Wailes, "Tide mills in England and Wales", *Jr Inst. of Eng., J. and Rec. of Transact.*, 1941, 51: 91–114.

marked incipient interest in France.[5] In Portugal, concern for tide mills goes back to the 1960s.[6] Nevertheless, the present interest in Europe only goes back to the mid 1970s[7] and especially to the eighties. Thus, not until 1988 did Jean-Louis Boithias and Antoine de la Vernhe bring out their remarkable book on the mills of Brittany.[8]

The tide mill of bygone times has inspired engineers' dreams and after lengthy and varied tribulations became the genitor of tidal power plants. To cater to local demands for power was the idea behind the tide mills that once dotted the coasts of England, Wales, France, Portugal, Spain, Belgium, The Netherlands, Germany, Denmark, Canada, the United States and China. Many are derelict today, some disappeared without a leaving a trace and efforts to save this industrial archaeologic patrimony are still timid. There are no molinological societies for the preservation—and restoration—of tide mills, such as exist for wind- and water-mills which are the ones that displaced them.

If the construction of France's Rance River tidal power plant sealed the doom of several tide mills still in a state of relative activity in 1966, an interest has recently developed, particularly in France and the United Kingdom, in rehabilitating tide mills and even putting some back to work; England, France and the United States have some "working" mills today. The "industrial" heritage of Iberic tide mill is also the subject of major salvaging and restoration efforts.

## 2.2 Tide Mills, Economics, Industry and Development

Tide mills have played an important role in the industrial and port development of Western Europe. These mills were sited on coasts, but also on rivers. The technique of tide mill building and utilization was exported to the Atlantic coasts of North America. Tide mills were gradually supplanted by the advent of newer technologies, though several remained functional, and at work, well after the end of World War II. They may be appropriately considered as forerunners of the modern concept of tidal power plants. A renewed interest in tide mills has been generated by industrial heritage historians and praiseworthy efforts at safeguarding and rehabilitation have blossomed, particularly in France and the United Kingdom; other countries are somewhat trailing. But is only history at stake?

---

[5] Philippe Veyrin, "Les moulins à marée du Pays basque", *Bull. Musée du Pays basque*, 1936, pp. 414–423.

[6] Xoaquin Lorenzo Fernández, "Muiños de mare", Oporto, *Trabalhos de Antrop. e Etnol.*, 1959, 17: 249–255; Fernando Castelo Branco, "A Plea for the study of tide mills in Portugal", *in* : *Ist Symposium of Molinology, Lisboa-Cascais*, 1965a, pp. 81–83; *idem*, "Moinhos de maré em Portugal", Lisboa, *Panorama*, 1965b, 4 (14).

[7] Claude Rivals, "Moulins à marée en France", *Trans. Third Symp. Intern. Molinological Soc.*, Arnhem, 1973.

[8] Jean-Louis Boithias & Antoine de la Vernhe, *Les moulins à mer et les anciens meuniers du littoral: Mouleurs, piqueurs, porteurs et moulageurs*, 1988, Métiers, Techniques et Artisans. Créer, 275 p.

Man has been awed by the sea, but not enough not to try to travel it, to fish or to hunt on it, to extract non-living resources, nor even to attempt to harness its power. He has indeed, for nearly a millennium, put to work the energy of the tides. And, after a rather long lapse of time during which the tide mill was considered as an archaic, artisanal method of providing mostly mechanical energy, he is again turning to them to tap a rather environmentally benign source of power. The tide mill of bygone times inspired engineers' dreams and after lengthy and varied tribulations became the genitor of tidal power plants.

Many sites are suitable to build what the French—the first to construct one—call *centrales marémotrices* but such plants are still few, except in China. Perhaps the Chinese have a more realistic approach: instead of thinking "big", such as with the Chausey Islands scheme in Brittany (France) or the multiple basins projects of Passamaquoddy (USA/Canada), they have thought of local needs which could be satisfied by small plants. To cater to local demands for power was the idea behind the tide mills that once dotted the coasts—and sometimes rivers—of principally England, Wales, France, Portugal, Spain, Belgium, The Netherlands, Germany, Denmark, Canada, the United States, and China.

An abundant literature provides a geographical panorama of the tide mills throughout the world.

Many tide or sea mills are derelict today and some disappeared without a leaving a trace. Efforts to save this industrial archaeological patrimony are still timid. Until recently no molinological societies vied for the preservation (and restoration) of tide mills, as exist for wind- and water-mills that often took their place. The construction of France's Rance River tidal power plant sealed the doom of several tide mills still relatively active in 1966, yet an interest recently developed, particularly in France and the United Kingdom, in rehabilitating tide mills and even putting some back to work. England, France and the United States have some "working" mills today. The "industrial" heritage of Iberic, and American, tide mills is also the subject of major salvaging and restoration efforts.

Is it too far fetched to suggest that the tide mill could function parallel to its impressive counterpart and provide in disinherited regions, in developing countries, in forlorn areas not in need of huge blocks of kilowatts, the power needed by modest local industries, albeit even if they are artisanal? Small can be beautiful, it was once said! It can also be useful and quite adequate. Why shouldn't tide mills like the Phoenix rise from their ashes?

## 2.3 Historical Development

### *2.3.1 The Middle Ages*

Tide mills are first mentioned in the Persian Gulf. In the 10th century the Arab geographer, Al-Magdisi Shams al-Din described the mills found at Bassora (Irak), on the Tigris-Euphrates delta, explaining how water turned the wheels as it flowed back to the sea.[9] In the 12th century, other mills started to be built along the Atlantic coast from Britain to the Basque Country.[10] Except for the Wooton mill (Hampshire), dating from 1132, England tide mills were built along the North Sea littoral: Bromley-by-Bow and Woodbridge (Suffolk) in 1135 and 1170, or Baynard's Castle near London in 1180.

From the 13th century on, mills were not only built along the East coast of England, (Wooton is the exception) but also along the southern coast at Southampton, Plymouth and on the Isle of Wight, even to the West in Wales, along the Severn estuary. Fifteen new mills were added in the course of that century.[11] Indeed, the number of mills seems to have increased steadily throughout the Middle Ages.

In The Netherlands the earliest known mill was built at Zuicksee in 1220. In France, the mill of Veulves (Normandy) was built in 1235 and those of l'Esbouc and la Nive (Basque Country) in 1251 and 1266.

On the Iberian Peninsula, the earliest mill might well be that of Castro Marim, built below the castle, on the right bank of the mouth of the Guadiana River. This mill is represented in a manuscript drawing of the fortress of Castro Marim in 1290.[12]

## 2.3.2 From 1492 to the End of the 18th Century

The Great Voyages of the 15th century, and especially the [re-]discovery of America by Columbus in 1492, undoubtedly affected the evolution of tide mills during the next century, especially near major ports such as Lisbon or Cadiz.[13]

In eastern Flanders, in Northern Brittany and in the Spanish Basque Country, mills functioned in these centuries while in the 17th and 18th centuries many tide mills appeared along the Atlantic coast due to the development of grain crops and the colonisation of America. For instance, some sixty new mills were established in Great Britain at this time.[14,15]

Credit goes to the French who, aided by the Micmac Indians, built in 1613 the first tide mill in North America; a double function mill, at Port Royal (Nova Scotia). Other mills soon followed in New England (Salem, Boston, Chelsea, Rhode Island

---

[9] Jean-Louis Boithias & Antoine de la Vernhe, p. 15.

[10] Rex Wailes, pp. 91–114.

[11] idem.

[12] Ernesto Veiga de Oliveira, Fernando Galhano & Benjamin Pereira, *Tecnologia tradicional portuguesa : Sistemas de moagem, Lisboa, Inst. Nac. Invest. Cient., Centro Est. Etnol., Col. Etnologia*, 1983, 2: p. 83.

[13] Ernesto Veiga de Oliveira et al., pp. 84–86.

[14] Walter E. Minchinton, pp. 339–353; Rex Wailes, pp. 91–114.

[15] Loïc Ménanteau, Eric Guillemot & Jean-René Vanney, *Mapa fisiográfico del litoral atlántico de Andalucía. M.F. 04 Rotala Barrosa (Bahía de Cádiz), M.F. 05 Cabo RocheEnsenada de Bolonia*, Junta de Andalucía & Casa de Velázquez, 1989, 2 maps at scale of 1: 50.000 + trilingual monography (58 p.).

Passamaquoddy Bay, Long Island).[16] In the 18th century more complex installations using tidal energy were constructed.

### 2.3.2.1 The 19th and 20th Centuries: Industrialisation and Abandon

Traditional mills continued to be built but other mills disappeared and later, electricity, gradually superseded, or complemented, tidal power.[17] Thus, turbines gradually took the place of hydraulics. In Britain, numerous mills closed down during the first half of the twentieth century.[18] Due to the privations brought by the Second World War throughout Europe, many mills continued to function; some into the sixties and seventies; Bauchet mill, for instance, did not close down until 1980. In Spain, some mills were still at work in the 1950s and in Portugal, six mills were still working in the 1960s.

## 2.4 Location of Tide Mills

To understand the extent to which tidal energy was used, one might first look at the geographical distribution of tide mills. Given the lack of data available for certain countries, however, the figures are, at best, approximate. Furthermore, this distribution does not take into account the type of mill or its construction date, let alone the fact that many have since disappeared. In all, over 800 mills were built on both sides of the Atlantic and the North Sea, and over half of these were on the European littoral. From Scotland to the Straits of Gibraltar, Great Britain, France, Spain, and Portugal once totalled some 500 mills! A closer look at their distribution, with particular attention to the areas of highest density, might prove informative.[19]

The most common location of tide mills is of course on coasts where tides have large amplitudes. Some have been placed on rivers, capturing flow of river energy, others on estuaries or bays, using the tidal currents. Often mentioned are those of the Danube and Tiber rivers. Some mills were reported to have been in use in the first century, and later in Asia Minor in the tenth p.ex. in Irak. Tide mills functioned on the Scheldt River in the sixteenth century. More recent are the river mills of Russia, Italy, and Hamburg (19th and 20th centuries). The Rance River lost many mills when the tidal power station was constructed in the 1960s.

---

[16] H. Creek, "Tidal mill near Boston", *Civil Engng*, 1978, 22: 840–841.
[17] Jacques Guillet, "Meuniers et moulins à marée du Morbihan", *Le Chasse-Marée. Histoire et ethnologie marine*, 1982, 5: p. 57; Bernard Le Nail, 26 p.; Jean-Louis Boithias & Antoine de la Vernhe, p. 15.
[18] Roger H. Charlier, 1982, pp. 55, 57 & 60.
[19] Roger H. Charlier, 1982, p. 61.

## 2.4.1 Spain

Mills dotted the coastline of the Asturias, Cantabria, Galicia, the Basque country, and Andalusia, with Galicia having the largest number highly concentrated in the Ria de Arousa. A new road linking Cantabria and Castilla, facilitating flour export to the Americas through the port of Santander fostered in the 16th century the construction of mills, many in and around Santander; the height of their activity was in the 17th century. Similarly salt-pan development in the 18th century encouraged construction of mills in the Bay of Cadiz; the first date back to the 13th and 15th centuries. Mills were at work in the Rio Tinto estuary. Minchinton (1988), Perez (1985), Lopez (1991), Cordon (1975) and Escaleza and Villegas (1985) have provided detailed studies of mills located in various provinces of Spain.

Some 250 mills were built from the Bay of Biscay to the Straits of Gibraltar. According to Luis Azurmendi[20], often all that survives of numerous mills is a significant toponym. The deep wide Galician rías along the western coast are ideal for the harnessing of tidal power. As elsewhere, the condition of Galician mills varies considerably from one site to another. In some cases the buildings and mechanisms are fairly well preserved, in others all that remains is a significant toponym or documentary evidence. As far as Aceñas de Burgo, in the Ría do Burgo is concerned, only 16th and 18th century documents can vouch for the existence of this early mill that by 1580 already required repairs.[21]

The highest concentration of all the European coastlines occurs in the Tagus estuary. Some 70 mills lined the littoral of the Gulf of Cadiz. Doubtlessly, mills were linked with the largest salt pans (6,000 ha) in Europe. The southernmost of the European Atlantic mills was in the Barbate marshes within the confines of the Straits of Gibraltar where the tidal range is already limited.

Besides tide mills on the Cantabrian and Galician coasts (Azurmendi 1985; Azurmendi et al. 1988), Spain was dotted, at one time, by such mills in Atlantic Andalusia. Remnants, or even derelict structures (at least 44) still mark the landscape from the mouth of the Guadiana River to the *Marismas* of Barbate. Eleven of these located in the *marismas* of the *rios* (rivers) Odiel and Tinto have been reported by Vanney and Menanteau (1985) et al. (1989). Of different types, they used to grind grain putting tidal energy to work to move millstones with paddlewheels. Several plans, the oldest of the Puerto Real area, date back to the 16th century and provide a fair idea of how these mills functioned.

On and near the French border, mills were built on rias, tidal channels, inside bays but also on islands, peninsulas, even beaches and salt pans.

While many mills have disappeared, surprisingly numerous derelict mills are virtually intact and could be easily restored. For example, the mills of San Antonio at Ayamonte or the Arillo on the double track road of San Fernando to Cadiz and yet, none has been made suitable for visitors. The English and Americans have done a better job in that regard.

---

[20] Luis Azurmendi Pérez, pp. 11–18.
[21] Report, Limia Varela & Benoît Bernard (c. 2, p. 2).

These mills are only a part of a hydraulic system containing retaining dykes and a pond that filled up at flood tide and emptied at ebb. Power was generated during ebb and in very rare instances were double-effect wheels part of the system. Restoration projects must thus involve the system and not be limited to the mill structure alone, as has been done occasionally (Brittany, France, for instance). The pond has been embanked in some cases, but in others it has been modified and is now used for different economic purposes, and in some instances well adapted to a new utilization such as aquaculture.

## 2.4.2 *France*

Things were not much better in France, at least until recent years. The Rance estuary (Ile-et-Vilaine), on the northern coast, with 17 mills[22] and the Gulf of Morbihan, on the south coast, with 19 mills, are the two areas with the highest density.

Brittany has some 100 of the 140 mills thus far catalogued on the coasts. The architecture of many of these mills has been distorted by transformation into private residences, restaurants and *crêpes* snack-bars. Public and administrative decision-takers have been made aware of the exceptional historical and archaeological value of these mills by a spate of books (Boithias and La Vernhe 1988) and articles (e.g. Guillet 1982, Boithias 1988). But some mills have been reconditioned: the Page family, who bought the Traou-Meur in Pleudaniel on the Trieux (Côte d'Armor department of France) has undertaken, since 1979, a remarkable restoration program encompassing site, building and mechanism, and made the mill "visitable". The municipalities of Trégastel and Perros-Guirec (also in Côte d'Armor) bought the Grand Traouieros mill and are proceeding to make it into a Museum of the Tide Mill.

A similar plan exists for the Derrien Rock Bridge (*Pont à la Roche Derrien*) mill in France's Finistère Department where an eco-museum of the *"teillage"* of flax is to be housed. Finally, the *Musée de la Cohue*, in Vannes, displays the reduced model of a tide mill and offers a video program dealing with those of the Morbihan Gulf. No traces are left of the two tide mills constructed at the entrance of the Bay of La Rochelle, near today's Chain Tower *(Tour à la Chaîne),* under a 1139 grant of Alienor of Aquitaine (Elenor). Mills were at work inside the fortifications of the Protestant stronghold of La Rochelle.

Though the majority of French mills were located in Brittany (Charlier 1982, 1993, Boithias 1988, Guillet 1982, Le Nail n.d.), others were sited in Normandy, Charente, Aquitaine and in the French part of the Basque country. Their golden age spans the 17th and 18th centuries. Basque mills were already at work in 1255 and 1266 (Le Nail n.d.; Veyrin 1936). A mill was constructed in Normandy in 1235. A lone mill in Dunkirk (French Flanders, described by Forest de Bélidor in the 1730s) was built at the end of the 17th century. It was connected to a wind mill and used both the ebb and flood currents, forerunner of sorts of the double effect tidal power

---

[22] Maud Bruneau, "Les moulins à marée de la Rance", *in*: *Bull. Féd. Fr. des Amis des Moulins,* 1982, 7.

plant. In Aquitaine mills functioned on the Nive and Adour rivers, in Bayonne, Ascain, Sopite and Saint Jean de Luz.

In France and elsewhere, port development played a major role in tide mill siting. Besides, tide mills were placed on boats anchored in the Danube (Romania) and Tiber (Italy) rivers; on the Loire (France) River, likewise, tide mills were mounted, in the Middle Ages on boats, apparently near Champtoceaux (France).

Abandonment of tide mills preceded the industrial revolution but intensified in pre World War II days and accelerated from 1950 through the 1960s. The "renaissance" of interest in tide mills is evidenced by the complete renovation of the rather derelict Birlot Mill on the Isle of Brehat (Homualk 1987) in 1977 and "re-inaugurated" in early 1998. The island was a favourite hideaway of Jules Verne and currently of Erik Orsenna (Arnouldt). Similarly, the mill in Arz is "back at work" since July 1998 (Foucher 1998).

## 2.4.3 Portugal

Mills were grinding in the Tagus estuary in the 14th century; over 100 worked grain near Lisbon, producing a.o. items, biscuits for the Portuguese navy. Across the river, 27 mills were at work in the Val-do-Zebro (18th century). Fifteen had been built in the 15th century in the Ria de Aveira. In 1900, 30 were reported near Minho (East Algarve (Montalverne 1991, Branco 1990).

Several mills have been reconditioned in Portugal (Montalverne 1991), mainly in the framework of a policy of natural spaces protection; such is the case for the mills of the *Reserva Natural do Tejo* and the *Parque Natural da Ria Formosa* (Ménanteau 1991).

## 2.4.4 British Isles

Tide mills operated mainly in England and Wales though some were located in Scotland and on the Isle of Wight. One is claimed (*Domesday Book*) to have existed already in 1,000 at the entrance to the port of Dover. Most mills, apart of the undated traces in Northwest Scotland, provided power to various "industries" and businesses in Suffolk, Essex, Sussex, Hampshire, Pembrokeshire, and London. Best sites are of course the North Sea coast and about the Severn River, proposed site of tidal power centrals. Among the earliest is the Bromley-on-Bow mill, built in 1135 and restored in the 14th century, now a London underground stop on the Heathrow Airport Line. Plymouth and Southampton had mills in the 14th and 15th centuries. Those of Southampton were further developed in the 17th and 18th centuries. Ewing Mill is currently a working museum and, incidentally, an economic success. Plymouth had a mill in the 18th century. Background information on these mills has been provided in papers a.o. by Minchinton (1977), Triggs (1989), Wailes (1941), Holt (1988), and Charlier (1982).

## 2.4 Location of Tide Mills

Of the 140 mills once found in Great Britain, many were along the North Sea coast, site of the earliest mills, and on the Thames estuary (e.g. Kingston). Along the southern coast, the highest density was to be observed near the major ports, such as Southampton[23], Plymouth and Portsmouth, and on the Isle of Wight, where the Bembridge mill was described by the grandson of the naturalist, Charles Darwin. In the West, St George's canal, the Severn estuary (Wales), Pembroke Bay, and the Isle of Man all had their fair share of mills. In fact, right along the British coasts, not omitting Northwest Scotland, there are numerous traces of tide mills, including late examples.[24]

### 2.4.5 Northern Europe

Far less is known about Northern Europe, nor do in depth studies seem to have been conducted in the Low Countries. Yet a mill is reported to have been at work in Denmark in the first century. A mill is reported in the 13th century in Zeeland and another in South Beveland. There were mills at Goes and in the canals on Walcheren Island's Flushings (Flessingue, Vlissingen). The last Dutch tide mills, according to a personal communication from Jacob De Waart of the [Dutch] Monological Society, would have disappeared during the Golden Age of the XIX Provinces (16th and 17th centuries). Belgian tide mills were river sited (Rupelmonde and Zwijndrecht) far upstream approximately 95 km from the Scheldt River mouth. The mills at the confluence of Rupel and Scheldt have given birth to a small regional museum.

### 2.4.6 The Far East

The authors were unable to get information on whether tide mills ever worked on the coast of Korea, a region whose high amplitude tides make it a preferred site for tidal power plants. Similarly, no tide mills were built in northeastern China. But, oral information garnered at the 6th International Congress on the History of Oceanography (Qingdao, China, August 1998) places numerous such mills on the estuaries, and coasts, of southeast China.

### 2.4.7 The Americas

Dutch and British settlers took their ideas with them and recreated on the other side of the Atlantic working environments, which they knew in the country of origin.

---

[23] J.P.M. Pannell, Old *Souththampton shores*, Southampton, David and Charles, 1967.

[24] Walter A. Minchinton & J. Perkins, *Tidemills of Devon and Cornwall*, 1971; Rex Wailes, *Tide mills* (parts I and II), London, Soc. Prot. Anc. Build., 1961; E. M. Gardner, *Tide mills* (part III), London, Soc. Prot. Anc. Build., 1963.

Tide mills are thus found in the New York area, such as the Van Wyck mill, or the Chelsea mill in Massachusetts. The Chelsea mill is back at work, restored in a US National Park. The Brooklyn Mill, built after plans by the Italian Veranzio, is one of the few still standing in the New York City area. A group of enthusiasts is at the origin of a revival movement in Maine. Bostonians mention a tide capturing station at the end of the nineteenth century falling victim to port extension. Likewise, French immigrants imported the tide mill technique to Canada. A *fac simile* of the Lequille mill has been erected near the tidal power station of Annapolis-Royal. Three hundred to 350 mills were built on the coastal stretch between Canada and Georgia (USA) with 150 in Maine and Massachusetts (USA).

On the shores of the Caribbean, several mills have been recorded (e.g. Surinam) and others undoubtedly existed both in the Lesser and Greater Antilles, where they were used for sugar cane processing.[25]

## 2.5 Distribution Factors

Several factors explain, to a certain extent, the geographic distribution of tide mills along the Atlantic littoral. These factors are linked to marine hydrology, the configuration of the coasts, and the development of ports. A sufficient tidal range, at least two metres, is the condition *sine qua non* for such a mill to function. There is usually a high concentration of mills in areas with a considerable tidal range (Rance Estuary, West coast of Scotland)

Nevertheless, tidal range does not appear to be a determining factor. In Spain and Portugal, there are high concentrations of tide mills despite a limited tidal range. A mill needs an indented coastline with inlets and small estuaries which can easily be blocked off by a causeway or marshes drained by numerous channels. Thus, rectilinear coastlines, whether rocky or alluvial, even if the tidal range is favorable, are not ideal places for tidal mills and most mills are found in estuaries or rias, on tidal channels, or within bays. Others, though fewer, are on islands or peninsulas on rocky coasts or on the beach itself, in a sheltered position in relation to the flow and the strong tidal currents.

In southern Brittany, numerous mills were built in the saltpans where they occasionally served to empty the reservoirs and ponds. Situated at the entrance to tidal creeks and channels taking water to the pans, the periodic emptying-out process served to clear away the deposits accumulated by the tidal flow, thus maintaining the depth of these channels (Isle de Ré, Charente coast, the rias, Bay of Cadiz).

Siting of mills is also related to port development. Numerous mills clustered around historic port cities such as London[26], Southampton, Plymouth, La Rochelle, Bayonne, Lisbon, Faro or Cadiz. Strategic and commercial reasons, linked to

---

[25] Jean-Louis Boithias & A de la Vernhe, p. 16.

[26] J. Boulton, *Neighbourhood and society: a London suburb in the sevententh century*, Cambridge, Univ. Press, 1987, p. 23; M. J. Power, "Shadwell: the development of a London suburban community in the seventeenth century", *London J.*, 1978, 4: 29–46.

supplying the population as well as the Navy, to the import and export of grains and flour led to their implantation within or close to the major ports. In certain cases, there were also technical factors. In La Rochelle, for example, the mills were used to clean the port using a complex system of sluices which allowed silt to be evacuated with the out-flowing water at low tide. The case of saltpans is similar.

Changes in the natural environment may also lead to abandon of a mill: sedimentation or anthropic ones (drying out of marshes, in-filling, *etc*).[27] When tides become insufficient to fill the pond, the abandoned tide mill is evidence of a changing coastal landscape.[28] Sometimes such mills could be dual powered during an interim phase, prior to their total abandon.

## 2.6 Mills and Their Environment

Considered here are tide mills with machinery that is moved solely by tidal power and, to a lesser extent, dual-powered mills. The latter depend on water from the river as well as on the tide and allow seawater to enter at low tide to keep turning the wheels. Several such mills dotted the Atlantic coasts. Double-effect mills use both incoming and outgoing tides.[29] The Thames mill provided part of the city's water supply from 1682 to 1849. Its 6-m water wheels installed under the arches of London Bridge, ran in either direction generating power with flood and ebb tides. This was also true of the East Greenwich mill[30], one Charente-Maritime, Captain Perse's Dunkerque mill, and a few others.

A tide mill is part of a unit that comprises the mill proper, outbuildings, a dyke and a pond. The workings of a classic mill are simple. At high tide seawater flows into a pond, protected by a dyke, through a sluice gate which closes automatically, under pressure from the water accumulated in the pond, as the tide begins to withdraw. The water flows out of the pond through one or several narrower gates and, in so doing, turns the hydraulic wheels, which may be overshot, undershot or midshot.

The energy provided by tides, though intermittent, is regular and inexhaustible because it is constantly renewed. A mill functions from three hours before to three hours after low tide. Since wind can influence the speed of the incoming tide there could be marginal differences in the times, but these rarely vary by more than half an hour each way. Thus, the miller, who had to wait for the ebb tide before he could set his grinding stones in motion, had to adapt his working hours to the rhythm of the cycles of the tide, though on average he could expect to work a total of 12.4 h in

---

[27] Loïc Ménanteau, *Zones humides du littoral de la Communauté européenne vues de l'espace / Wetlands of the European Community littoral seen since space / Zonas húmedas del litoral de la Comunidad europea vistas desde el espacio*. T.I France, España, Portugal, Italia del Nord. Junta de Andalucía, Casa de Velázquez, CNES & SPOT IMAGE, C.M.P.R., pp. 45–134.

[28] Loïc Ménanteau et al., 1989.

[29] R.H. Charlier, Tidal energy. Van Nostrand-Reinhold, New York, 1982

[30] M. Gregory, *A treatise of mechanics*, 4th ed., London, 1826.

every 24. The mill could not be used at all during neap tides. Where tidal range is limited a coefficient of at least 65–70 is required.

There are numerous variations on this pattern, depending on the geographical regions.

### 2.6.1 Dikes

Dikes are built along a single continuous line, which may be straight or curved depending on the nature of the coastline and the degree of exposure. Their length varies considerably, though usually it ranges from 100 to 250 metres. Dikes can also be considerably longer (e.g. 300 m Spain and 450 m in Brittany) or shorter. They may also be built in sections taking advantage of rocky alignments and pebble spits forming part of a natural dike (on the Ile de Bréhat, Brittany and on the Illa de Arousa, Galicia).

Variations in width are also considerable, though three to five metres is about average. The width of a dike depends on the resistance to be opposed to the sea, the strength of the tides as well as on the shape of the dyke itself and the methods used in its construction, e.g. Cosquer mill (Finistère, Brittany) 18–23 m. Galician dikes can be no more than a metre wide. Often, the dyke is merely an earth levee protected by dry stonewalls. In some cases, the dyke is a thick wall strengthened by buttresses. Periodic de-silting of the pond allows the miller to raise or widen the dyke using the silt removed. Silting today adversely affects many of the small Chinese tidal power plants.

### 2.6.2 Entrance Sluice Gates

Dykes have one or more openings formed by automatic, hinged gates through which seawater flows in at high tide. These "entrance" gates, usually three to four metres wide, may be found close to the mill or several dozen metres away from it, depending on the mill's position on the dike. Occasionally, they are right under the mill. The Molino del Arillo in the Bay of Cadiz is also unusual in that its entrance sluices are placed perpendicular to the two wings of the building with a double gate under the central section of the main wing.

Sometimes they are divided into two or three sluices or watergates separated by solid, stone, internal dividing-walls. In Cantabria[31] and in Galicia[32], most mills are equipped with double gates. The separating wall looks rather like a ship's prow stretching into the sea, which facilitates the flow of water into the pond at high tide. Contrary to Brittany, where there is usually only one entrance gate per mill,

---

[31] Luis Azurmendi Pérez, p. 25.
[32] Begoña Bas López,1991, p. 111.

2.6 Mills and Their Environment

Spanish and Portuguese mills often have several. The gates fixed to these sluices, called "gates to the sea" (*portes de la mer*) in France, are of two types and move either on a horizontal or a vertical axis. The former appear to be more frequent on the southern European coasts, though the Eling (U.K.) tide mill[33], constructed on a toll bridge, had flap valves. In the latter, i.e. the gates moving on a vertical axis rather like shutters, a stone fixed to the inner wall of the dyke limits the aperture to an angle of 90°. There are also combinations of sluice gates that are both automatic and manual.[34] In such cases the lower part is on a horizontal axis while the upper part is fixed and slides up and down between the uprights rather like a sash-window. When the sluices are fixed and sliding, the mill is in fact a water mill, which only rarely uses tidal power, otherwise it would require constant attention from the miller who would have to intervene with every changing tide.

### 2.6.3 Ponds

The size of ponds varies considerably since it depends on the coastline, the position of the dike and on the tidal range. In Brittany, on the whole, ponds are larger in the South (8 ha in the Gulf of Morbihan) where the tidal range is more limited than in the North (2.9 ha) where the tidal range is greater.[35] The volume of water stored determines the mill's energy capacity and the number of hydraulic wheels that can be activated. In 1826, engineers recommended that the pond of the Herrera mill (Bay of Cadiz) be enlarged so as to increase the number of grinding stones. In Brittany ponds range from 0.2 ha (Combrit mill) to 30 ha (Ludré mill). The size of the Noyalo mill pond (77 ha) in the Gulf of Morbihan is exceptional since this is a dual-powered mill. Indeed, while some milponds range from 1.5 to 2 ha, some are 3–6 ha large and others go from 8 to 13 ha.

Some mills have several communicating ponds thereby providing a longer working time for the mill, about 15 h in all, as water flows from one pond to the next. In France, the only known example is that of Pont-Canada (Côtes d'Armor), but several examples are found in Great Britain (Bishopston and Falmouth mills, in Sussex and Cornwall respectively). In the Bishopston mill, an external channel linked the second to the first pond. Water from streams flowing into the ponds can help to fill these, but will rarely suffice to turn the wheels.[36]

---

[33] William Smith, *The tide mill at Eling. History of the working mill*, Southampton, Ensign Publications, 1989; M. Southgate, *The old tide mill at Eling*, Southampton, Eling Tide Mill Trust Ltd, 1991.

[34] Jean-Louis Boithias & Antoine de la Vernhe, pp.155–156.

[35] Jean-Louis Boithias & Antoine de la Vernhe, p. 144.

[36] Walter E. Minchinton, pp. 339–353; *idem*, "Moulins à marée: étude préliminaire", *L'onde* (Rev. de l'Assoc. des Meuniers d'eau), 1979, pp. 1–10; *idem*, "History to day", in: *Power from the sea*, London, 1980, 30: 42–46.

## 2.6.4 Exit Gates

The purpose of the exit gates in single effect mills is to direct water accumulated in the pond during the flood tide towards the hydraulic wheels during the ebb tide. Their number depends on how many wheels are to be turned, though there are never more than three or four in Northern Europe. The channels are as wide as the wheels: 0.50 m in Southern Brittany and one metre in the North.

Further south and throughout the Iberian Peninsula the narrow, funnel-shaped channels give greater force to the water falling on the horizontal wheels. Their number is usually greater than in Northern Europe and never less than three. In fact, in Spain, mills with six or more channels are common. The Cantabria Santa Olaja mill has nine, Joyel and Castellanos have eight each. In the Bay of Cadiz, the San José mill has eight channels. In Galicia, their number varies from three to six, with narrow channels being associated to wider ones.[37]

## 2.6.5 Wheels

Another significant difference between Northern and Southern European tide mills: while the former are generally equipped with vertical wheels, whether internal or external, the latter tend to have horizontal, internal wheels. A census carried out in Brittany in 1809 showed that 80% of the 6,450 wheels were vertical, and only 20% horizontal[38], the latter being extremely rare in southern Brittany. While in Great Britain almost all vertical wheels are external, across the Channel, in Brittany, they may also be internal.

The number of wheels varies from one to three whether they be internal, external or a combination of the two, as is the case in the Pencastel mill (Morbihan). They may be on the same course, parallel, off-course or placed one after the other. Waterwheels in tide mills were usually wooden, though some were spoked wheels with iron hubs. The wheel shafts, of wood and iron, also varied considerably in size.

The wheels of the Rhode Island mill (USA) were eight metres wide, with a diameter of 3.35 m and weighed 20 tons.[39] In England, Stambridge Mill, on the river Roach, near Rochford, had a wheel 9.9 m wide with a diameter of approximately 5.5 m. Mounted on a wooden shaft it drove three pairs of stones. The dual-powered (water and tide) mill at Demi-Ville (Morbihan) had wheels with a diameter of 5.60 m. On the whole wheels tend to be narrower in the South than in the North.

In Spain, mills tend to have a far greater number of wheels. The mills of La Venera (Cantabria) and El Arillo (Bay of Cadiz) had thirteen and twelve respectively,

---

[37] Begoña Bas López, 1991, p. 113.

[38] Jean-Louis Boithias & Antoine de la Vernhe, p. 186.

[39] Roger H. Charlier, 1982, p. 63.

though in 1828 engineers suggested equipping the latter with as many as 20 wheels.[40] Each of these wheels is placed at the bottom of a kind of individual cylindrical stone well at the base of which there is a wooden beam into which the shaft is inserted. The shaft transmits power to the grinding stones.

These horizontal wheels are of two types: "a rodicio" and "a rodete". The former, or "ruedas a rodicio", are older and have curved paddles. Their diameter varies from 0.90 m for the Fontas mill in Lagos to 1.50 m in the A Mouriscas mill on the Sado estuary (Portugal). The wheels of the A Mouriscas mill, with their 22 curved paddles, can reach speeds of 150 rpm.[41] The "rueda a rodete" does not have paddles, but a smooth external surface and straight, solid wood spokes. In many cases, this newer type of wheel replaced paddled wheels since, being placed lower on the supporting beam, it prolonged the working time of a mill.

## 2.6.6 The Mill

The mill itself may be built in the middle or at either end of the dike, the exact place being usually determined by the topography and/or hydrology. Industrial mills, however, are usually built at one end of the causeway as this offers easier access.

The architecture of tide mills is largely comparable to that of water mills. From the mid-nineteenth century industrial mills could have three, four and even five storeys. The floor plan can be up to three times larger than that of the traditional mills they often replaced (e.g. Rochegoude mill, from 8.30 m × 10 m to 13 m × 20.50 m).[42] These traditional mills show considerable variation in size. Most of them are of a more or less pronounced rectangular shape (e.g. in Galicia, 9 × 13.70 m for the De Cura mill, 6.20 m × 20.50 m for the Acea da Ma mill).[43] Although most are one or two storeys buildings, in the South a third floor may be added for the miller's dwelling. From the 19th century, in Brittany, the miller's dwelling tends to be built away from the mill proper.

In mills equipped with horizontal internal wheels, the cavity under the building is reserved for these and for the mechanisms transmitting power to the grinding stones in all cases. When wheels are vertical, this cavity is virtually empty (e.g. Rance estuary).

---

[40] Report, Limia Varela & Benoît Bernard, *Grande reconnaissance militaire de l'île de Léon*, San Fernando, 1826, Arch. Serv. Génie, Serv. Hist. Armée de Terre, Cadix (c. 2, p. 2).

[41] Ernesto Veiga de Oliveira et al., pp. 133–134.

[42] cf. fn. 41.

[43] Begoña Bas López, 1991, p. 112.

Jean-Louis Boithias[44] identified several architectural types:

(a) the upper part containing the grinding wheels is a light wooden structure, built on piles above solid stone foundations (e.g. Vauban in Paimpol, Lupin in the cove of Rotheneuf);
(b) the side walls are left open and blocked in with wooden planks (e.g. Traou-Meur, Grand Traouéiros, in Côtes d'Armor);
(c) in simple masonry (e.g. Birlo, Dourduff-en-Terre).

Buttresses, whether on the front or sides, sometimes reinforce the walls providing stability and greater resistance to storms, which can be frequent in the area, and act as breakwaters.

Mills situated south of Brittany, from the coast of Charente-Maritime all the way to the Bay of Cadiz going through the Basque Country, Galicia, and the various Portuguese regions, all presented the same aspect on the sea-facing facade: a succession of arches which, when their number is high, gives the mill the appearance of a covered bridge. There are fewer arches than there are exit sluices or hydraulic wheels. Thus, in Cantabria, the Santa Olaja mill has seven arches to nine wheels, that of La Victoria three arches to eight wheels and La Venera six arches to thirteen wheels.[45]

In Portugal and in the Gulf of Cadiz mills are often of a considerable size and together with their numerous outbuildings frequently constitute veritable complexes. Indeed, some mills had their own chapel. Huge pottery urns were used for storing fresh water (e.g. San José mill, Bay of Cadiz).[46] Above the machinery and the grinding stones there were grain stores. In the upper part of the mill, whether in the area reserved for milling or in the store rooms above, most mills had a door through which the sacks of flour could be loaded directly on to boats at high tide.

On the whole, tide mills were used for grinding grains (wheat, barley, rye and, in Brittany, buckwheat) and complemented windmills.[47] Between 1823 and 1828 the Cadiz mills ground grain to feed the French occupation forces. Tide mills were also put to other uses. As already mentioned, the mill on London Bridge provided fresh water for the City up to the late 19th century. Bidston, in Cheshire, was an iron-slitting mill and some mills were used for rasping and chipping dyeing woods and the manufacture of tobacco stems. Some mills were providing power for breweries. The copper smelting works established in the Hayle mill in Cornwall in the mid-eighteenth century were eventually abandoned as no one could be found to work when the tides provided power at different times of day and night. Slade's Spice Mill, in Chelsea, Massachusetts, was used, as its name indicates, for the grinding of spices. The two tide mills operating in Passamaquoddy Bay prior to 1,800, however, were standard grist mills.

---

[44] Jean-Louis Boithias & Antoine de la Vernhe, p. 164.
[45] Luis Azurmendi Pérez, pp. 43–48 & 69.
[46] Report, Limia Varela & Benoît Bernard (c.2, p.2)
[47] A. Gutierrez, J. Muñoz & S. Ariztondo, *La industria molinera en Vizcaya en el siglo XVIII*, Bilbao, Univ. de Deusto, 1984.

In Trégor (Brittany), several early mills were transformed and used for flax. In the Spanish Basque Country, the Mallukisa mill in Murrieta (Province of Vizcaya) was used to grind kaolin that was then sold to the San Mamés porcelain factory. Some mills situated within saltpans were used for washing salt as was the case of the Loix mill (Ile de Ré, France) in 1824–1825, while others were converted to paper mills, saw mills and carpenters' workshops and even used for the manufacture of ice. Still other mills were used in drying up of polder (Netherlands).

## 2.7 Renaissance of the Tide Mill?

The use of the energy generated by tides[1] is thus an integral part of the history of the sea. The recent interest in sources of renewable energy has undoubtedly contributed to the present attention given to the forerunners of tidal power stations.[2] The historic value of tide mills is being recognised, perhaps buttressed by possibilities of reviving their use towards modern versions. Changing attitudes towards industrial archaeology[3] and growing concern for our maritime architectural and environmental heritage have also influenced the present trend towards the study and conservation of these remains, proof of the ingenuity of our forebears.

After they stopped grinding wheat and other grains, most of the mills were either abandoned or simply pulled down. All too often their environment (dykes, ponds) was also significantly altered.

Mill dikes, when not partly or totally destroyed, have been cemented, raised or widened in order to make way for new roads. They can also serve for the construction of houses. The most serious problems are due to the deterioration of millponds. Since they are no longer filled and emptied with every tide they have a tendency to naturally silt up. Silt and vegetation soon invade a large sector of their surface area. This natural trend, resulting from the fact that the pond is no longer cleared with ebb tide as it was when the mill was functioning, is further aggravated when the ponds are cut by other dikes or roads.

Man is also responsible for the drying up or filling in of several ponds for agricultural purposes or to turn them into building land. In Cantabria, pressure from developers is very strong indeed. In Huelva, it is industrial waste (calcium phosphates, *etc.*) that covers the surface of several former millponds on the right bank of the Tinto estuary[48].

Ponds are often also used as fresh water reservoirs. Nevertheless, the most frequent use is for aquaculture as happens in Brittany and Cantabria. In Morbihan (France), some former millers have turned to oyster farming, using their millponds for the breeding of oysters. In the same French department, there are plans to create several oyster nurseries in such structures. This is, on the whole, positive as it helps to maintain the pond's physiognomy, but can become a threat when it is a question

---

[48] Jean-René Vanney & Loïc Ménanteau, M.F. 02.

of developing aquaculture without taking into account existing structures, as has happened in several parts of the Iberian peninsula (Bay of Cadiz).

It is an aberration and a contradiction that the Rance Tidal Power Station is the direct cause of the silting up of 17 millponds stream up on the estuary. The power station has reduced the tidal range and the currents within the estuary.

As pointed out earlier, a certain number of mills have been saved from destruction or total ruin by being converted into dwellings, antique shops, restaurants and other uses. This favored their conservation and the restoration work is often of the highest quality: in Brittany Rochegaude or Buguélès on the northern coast and Coët-Courzo, Le Bono, Mériadec, Pomper, Hézo and Pencastel on the Gulf of Morbihan. Unfortunately, such conversions often lead to the total disappearance of the inner machinery of the mills, while the hydraulic wheels fall into disuse and gradually rot.

And, as also said already, sometimes, by being restored or turned into museums mills become tourist and cultural attractions. In Great Britain, the Woodbridge tide mill (Suffolk) was restored between 1953 and 1970. The Eling Tide Mill Trust acquired the mill on Southampton water in 1975 and had it restored by 1980 when it became a working museum attracting large numbers of visitors.[49] The Beaulieu mill was also restored and was in use in 1970. The Ruppelmonde mill in West Flanders (Belgium) now houses a milling museum. The restored Traou-Meur mill (Brittany), opened to the public in 1989, the Petit Traouiéros mill houses a permanent exhibition on tide mills. There are others examples and plans in France, in Spain, in Portugal.

## 2.8 A Preservation-Worthy Heritage

Industrial history, even archaeology has gained *droit de cité*. Mills are a non neglected folkloric but also cultural, even scientific topic, embracing wind, water (river) and tide mills. The latter have gotten more attention, apparently, from study groups aiming at the preservation of old buildings, engineers and historians than from molinological associations. But interest is growing.

Coastal protection has come from medieval industry as well. The demand for mechanical power fostered tapping of ocean energy. Development of tide mills particularly from the 15th century on encouraged their better protection against severe storms. Often, indeed mills were completely destroyed by the sea, a consideration that in fact led to their demise in various sites (e.g. England). Millers protected their structures seaside with abutments which constituted strong breakwaters and, indirectly, benefited neighboring areas as well. Some of these structures have been conserved.

---

[49] W. Smith.

## 2.9 Conclusion

As they fell in disuse, tide mills were abandoned or torn down and their appurtenances, such as dykes and ponds, altered and often put to other uses. Their buildings served as quarries for construction material, or made room for villas or "council flats". The original architecture is commonly affected by additions in other materials. The dykes made way for new roads. Millponds have frequently silted up and vegetation invaded them.

Some ponds are now used as freshwater reservoirs, but more often have become aquaculture basins and some millers have turned into oyster farmers. This has, in many instances, saved the physiognomy when the existing structures are taken into account when expanding the facilities.

Mills closed down in France between 1950 and about 1967. Their end started in England and Wales around 1900, a few survived World War II but were abandoned thereafter or converted to other uses by 1950. Italian prisoners of war worked at tide mills near Southampton. The Spanish mills became derelict between 1936 and the early fifties.

These true ancestors of the contemporary electricity generating tidal power plant could perhaps find a new life as a small mill serving an immediate community, an idea perhaps reflected in the many small tidal power plants functioning in China. (Zu-Tian 1989) The question may be appropriately raised, it seems, whether tide mills, or their remnants, ought to be preserved as a rare heritage, monuments of bygone industrial technology, or considered a technology That, with refinements, could be usefully resurrected and put back to work. Some have been rescued from destruction by being transformed into restaurants, antique shops and even dwellings, others are musea, even working ones.

**Fig. 2.1** Density distribution of relicts of tide mills in Western Europe. (L. Ménanteau)
*Legend*: A = Anglesey; AR = Ria de Arousa; B = Bayonne; C = Cheshire; CA = Coast of Armor//CA (in Spain) = Cadiz; CH = Cornwall/CH (in France) = Charente Maritime; D = Donostia (a.k.a San Sebastian); E = Essex; F = Finistere; G = Huelva; K = Kingston; L = London/L (in Portugal) = Lisbon and Tagus R. mouth; LA = Loire Atlantique; M = Morbihan Gulf; P = Pembroke; PL = Plymouth; R = Rance; S = Southampton; SA = Sado Estuary; SF = Suffolk; SG = Gibraltar (Straits of); SU = Sussex; W = Wight (Isle)

**Fig. 2.2** East Medina mill, Wippingham, Isle of Wight, (Rex Wailes, *Tide Mills in England and Wales*, 1940)

**Fig. 2.3** Carew, Pembrokeshire, tide mill on Carew River. (Rex Wailes, *Tide Mills in England and Wales*, 1940)

**Fig. 2.4** Pembroke, Pembrokeshire, tide mill on Pembroke River. (Rex Wailes, *Tide Mills in England and Wales*, 1940)

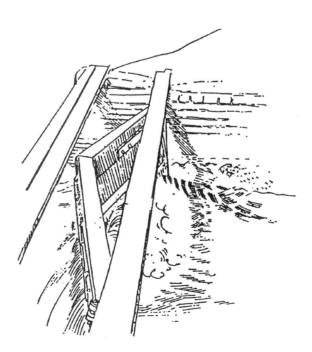

**Fig. 2.5** Sluice gate of Birdham tide mill. (Sussex, U.K.) (Rex Wailes, *Tide Mills in England and Wales*, 1940)

**Fig. 2.6** St. Osyth tide mill, Essex. Stones, wheat cleaner, and sack hoist

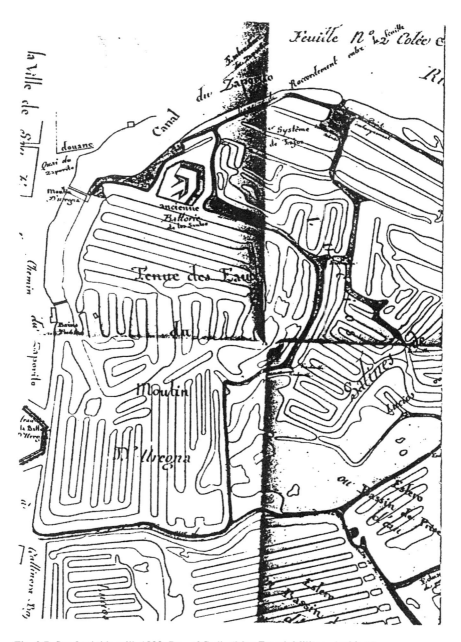

**Fig. 2.7** San José tide mill. 1823. Bay of Cadiz (Map French Military Archives)

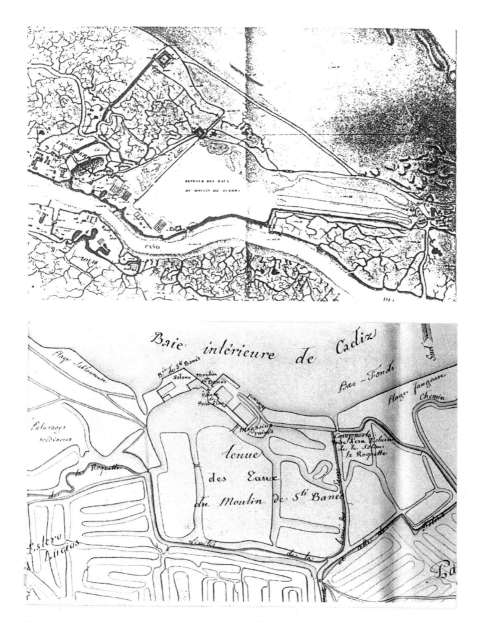

**Fig. 2.8** (**a**) Bay of Cadiz (**b**) St. Banes tide mill, 1823

**Fig. 2.9** Tide mill machinery as pictured on a 1703 engraving

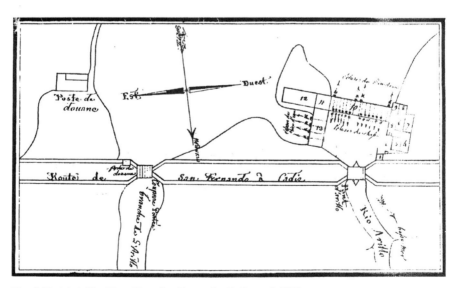

**Fig. 2.10** (**a**) Arillo tide mill on San Fernando-Cadiz road, 1823

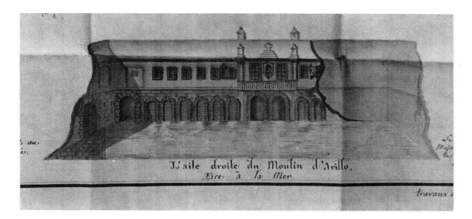

**Fig. 2.10** (**b**) Arillo tide mill located on road to Cadiz, new facing the sea. (figs. 2.10 belong to Archives of French Land-forces, now in Vincennes, France) (**c**) Present condition Arillo tide mill. (Photo L. Ménanteau)

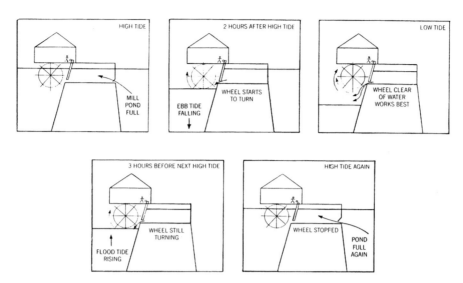

**Fig. 2.11** Operation of a medieval tide mill

**Fig. 2.12** Eling mill near Southampton (Engl.) Operating reconstructed tide mill (bakery and museum). (Drawing by Mel Wright)

**Fig. 2.13** Eling mill, restored 1980 (Ph. D. Plunkett)

**Fig. 2.14** Traou-Meurmill, Côtes d'Armour (Ph. L. Ménanteau)

**Fig. 2.15** Grand Traouiéros mill and dike, Tregastel, Côtes d'Armour (Ph. L. Ménanteau)

**Fig. 2.16** Uregna tide mill (20$^{th}$-century) on Zaporito mole (arch. post-card)

**Fig. 2.17** Arillo tide mill, Cadiz (Ph. L. Ménanteau)

**Fig. 2.18** Bartivas tide mill near Chicanadela Frontera. (Photo L. Ménanteau)

**Fig. 2.19** Bartivas (Ph. L. Ménanteau)

**Fig. 2.20** Mériadec mill, Badens (Morbihan). (Photo L. Ménanteau)

**Fig. 2.21** Ancillo mill, Santoñary, Cantabria (Photo Azurmendi)

**Fig. 2.22** Keroilio mill, Plougoumelen, Morbihan (Ph. L. Ménanteau)

**Fig. 2.23** Petit Traouiéros mill, Perros-Guirec, Côtes d'Armor (Ph. L Ménanteau)

**Fig. 2.24** 17th-century Pen Castel mill, Arzon, Morbihan (Ph. L. Ménanteau)

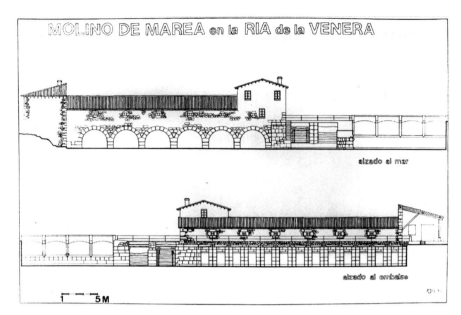

**Fig. 2.25** Tide mill on the Venera Ria

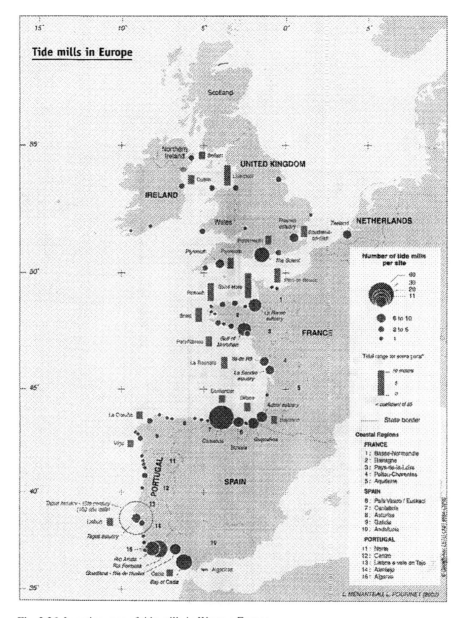

**Fig. 2.26** Location map of tide mills in Western Europe

# Chapter 3
# The Riddle of the Tides

## 3.1 What is a Tide?

The word tide refers to the rise and fall of sea level relative to the land. Such vertical movements differ considerably according to seasons, astral alignment, and other factors, including geomorphology. The movement is produced by the gravitational attraction, the "pull" of several celestial bodies, but overwhelmingly by that of moon and sun. The "pull" of the moon is the strongest because, although of smaller mass than the sun, it is far closer to earth.

In popular thought tides affect the sea, yet, there are tides in lakes and in the atmosphere, even within the solid crust of the earth. Besides the astronomical factors, the coastal geomorphology, coastal water depth and ocean floor topography play a non-negligible role in local tidal situations. Meteorological conditions have an influence. Proposals to take advantage of geomorphology to harness tides' energy have been put forth.

There is also a to and fro movement that accompanies tides. Such horizontal movement is referred to as *tidal current* and has been considered as a source of energy for electricity generation. Knowledge about tides has many uses but in this volume it is the energy dissipated by tides and the force of the tidal current that concerns us.

The designation of "tides" has also been used in connection with phenomena that are neither cyclic nor predictable. So-called "meteorological tides" are water level variations caused by winds and barometric pressure changes. Persistent strong winds bring about the piling up of water along coasts, reaching sometimes several meters in height, turning into hurricanes and violent storms.

## 3.2 Types of Tides

Various alignments and combinations of the gravitational forces acting and the respective relative distances between moon, sun and earth influence greatly range and amplitude of tides. So does the angular positioning of sun and moon. Their relative speeds, inclination of apparent paths of moon and sun in respect to the celestial equator, and finally the moon's and sun's length of arc of movement across the sky, affect the length of time during which "augmented" tides exist.

The daily lunar retardation is a factor that has played a primordial role in tidal power utilization.

Tides have been called according to their recurrence phase within a daily cycle. Those that display one high and one low in a tidal day are *diurnal* tides. That type is encountered, for instance, on the coasts of Alaska and Korea. The lunar day differs, of course, from the earth's day; the mean tidal day may be considered synonymous to the mean lunar day and of similar length. It may differ where considerable deviations in such length occur, depending on specific astronomical conditions. More precisely the tidal day is the period of time between the larger maxima or minima of two same type tides. The length of the mean lunar day, or period of time separating two successive upper transits of the moon across a local meridian, at the equator, is 24 h 50 min and 24.9 s.

Tides are *semidiurnal* when in a tidal day two high and two low tides occur. They are encountered along the Atlantic coasts of Europe and the United States. There must be only a minimal inequality between successive tides of the same phase.

As one can surmise, there exist also *mixed* tides that hold characteristics of both diurnal and semidiurnal tides. Mixed tides are found along the Pacific Ocean coasts on the United States. Their own characteristic is the large diurnal inequality. Finally an *agger* [or *double* tide] has either two equal height maxima and a single small minimum, or two separate minima separated by a single small maximum.

Oceanographers speak of a *vanishing* tide when tide levels merge, tidal days overlap and there occurs, except for one tidal phase, disappearance.

To these three major types several others should be added; these are designated by lunar, solar or luni-solar positions. The lunar tide-raising force increases as the moon nears a position in the zenith. In low latitude regions larger range and greater amplitude tides are produced. These are called *tropic tides*. Similarly the moon and sun combined forces at syzygies near the solstices increase diurnal inequality, range and amplitude, leading to *solsticial* tides.

The moon crosses the ecliptic bi-monthly. The sun crosses the celestial equator at each equinox. When thus both lunar and solar semidiurnal tides component are increased the consequent tides are *equinoctial;* they are named *equinoctial spring* tides occur when the preceding situation when a syzygy alignment also occur. Such alignment has earth, moon and sun on a straight line. Still further increased tides are then observed: these *spring* tides "spring up". There is no relationship with Spring as a season. Spring tides occur at new and full moon. Their "opposites" are *neap* tides; these subdued tides occur at the quadratures.[1] (Wood 2003).

*Equatorial* tides occur bi-monthly each time the moon crosses the equator. The time between successive perigees of the moon is its anomalistic period; *anomalistic* tides are closely related to the moon's anomalistic period. Fluctuation of tidal range then occurs. Such tides are observed in the Bay of Fundy (Canada). Very close to perigee—a day or so—*perigean* tides occur that are also higher than usual. When

---

[1] First and third quarter-moons.

perigee and syzygy are very close to each other the more powerful tides are called *perigean spring* tides. These tides can be upgraded to *proxigean spring* tides. Finally the effect of the lunar parallax may lead to the occurrence of *extreme proxigean spring tides*. Potential energy of tidal power sites has been calculated by Bernshtein (1965), Gibrat (1964) and Mosonyi (1963) leading to the mathematical expression

$$E_p = K \cdot 10^6 AR^2$$

in which the coefficient K is the only to vary (1.92 and 1.97). R is either the average range (in meters) or the average range of the equinoctial tide, A is the basin area (in km$^2$), and $E_p$ the potential energy (in kW/h /year). A type of utilization factor is introduced which is based on the system selected : for a single or double basin plant and a single tide it is 0.224, double tide 0.34 but 0.21 for a double basin ; for a double basin, single tide with reverse pumping it is 0.277 and with pumps 0.234.

## 3.3 Tide and Tidal Current

Tides and tidal currents must be differentiated, and their relation is not the same at all sites : the first are the vertical movement of the sea level (rise and fall), the second means the horizontal flow, sometimes called ≪ run ≫. Tidal current power has, besides in Iceland, also been harnessed in the Faroë Islands. Information on plant and years of operation proved to be impossible to obtain. Even Icelandic specialists were unable to provide specifics at a US-embassy sponsored conference in 1972.

The current experienced at any given time is usually a combination of tidal and non-tidal currents. The tidal current in the unimpeded open ocean is normally rotary, clockwise in the northern and counter-clockwise in the southern hemisphere. Speed varies, with two maxima and two minima. It is reversing where flow—p.ex. straits or rivers—is restricted. The movement away from shore (downstream) is the ebb, and toward shore (upstream) is the flood. The maximum speed is attained at high and low tide, with directions being opposite to one another; it is nil when the direction inversion occurs.

The expression *tidal current* may lead to some misinterpretation as one lives and works on one side of the Atlantic or on the other. British oceanographers reserve exclusively the term current to the non-tide generated horizontal flow of water. A "current" resulting from tidal action is, in Anglo-Saxon usage, a *tidal stream*.

## 3.4 Tide Generation

The tide-generating forces encompass the gravitational pull of sun and moon and the rotational force of the earth. The difference in tide occurrence is caused by the *tidal day* being 1.035 times as long as the *solar day*. Superimposed are the earth's revolution and the moon's declination. The tidal effects caused by other celestial bodies are all but negligible due to their distance from the earth. These forces create high tides on the sides of the earth nearest to and farthest from the moon, separated by a low tide belt.

Most tides occur twice a day (two highs, two lows) but in some geographical areas there is only one tide a day. The periodicity of other types of tides may be far less, for instance that of equinoctial and spring tides, or even exceptional such as those occurring less than once a century (e.g. March 27, 1967 in the English Channel).

Simple theory fails to accurately describe the tidal phenomenon because the earth is not uniformly covered by water, water depth varies, the Coriolis deflection plays a role, and so do geometrical shapes of basin and coast, and basin- or bay-size.

### 3.4.1 Power Generation

Theoretically the rise and fall of the tides dissipates 3,000 million kW of which one billion in shallow seas. Four fifths of the theoretically available energy was eliminated because of too small a head, engineering problems with the selected site and/or distance to a market. The situation changed dramatically when low-head turbines were developed and when various systems were proposed to re-time production by adding storage basins or by judiciously choosing opening and closing times of tide gates and thereby synchronizing power generation with peak demand.

Yet, one would dare talk of an eventual Kafkaesque situation: the rentability of a station depends, at least sizeably, upon the size of the retaining basin. Lower production costs can be obtained with larger basins and larger tidal ranges. But large basins require long dams thus a bigger capital investment, though by doubling the dam's length basin size is quadrupled!

Gibrat (1956) devised a mathematical site formula based on dam length (L) and natural energy (NE) being the annual theoretical production (in kWh) of unit [bulb-]turbines operating in both ebb and flood directions. This leads to a coefficient k that the smaller its value, the better suited is the site:

$$k = L/NE$$

The basin's geometric shape and surface will, of course, determine the complexity of barrage building (Rouzé 1959). The ratio of the aperture section to the basin's surface is important : its dimension in absolute value, basin minimum depth, basin opening and more or less widening shape.

Briefly put, there are oceanographic, geographic, geologic, technologic and economic factors at play, not to mention environmental, and ecologic, considerations.

The potential energy of tidal movement, for a 6.2 h tidal cycle, can be estimated approximatively by the formula

$$E = A' \, mg \, I \int_0^H h \, dh / T = 1/2(103.10^3)H^2(A \cdot 10^6)/60.60.62 = 225 \, AH \; [kW]$$

## 3.4 Tide Generation

Wherein

A = basin surface area in km$^2$
A'= basin surface area in m$^2$
g = gravity acceleration (m$^2$/s)
h = tidal range (m), head (m)
m = mass of seawater

To visualize the possibility of harnessing tides' energy the following formula provides the incremental amount of energy per cycle obtainable from the water displacement

$$\partial E = gR(DS\_R) \tag{3.1}$$

integration gives

$$E = DgS \cdot R^2/2 \tag{3.2}$$

and with « double effect operation » the formula becomes

$$E_{max} = Dg \cdot R^2 S \tag{3.3}$$

wherein R is the tidal range, D water density, g acceleration due to gravity, S the area of the enclosed basin.

However only about 30% can in fact be retrieved, hence for a basin of one square kilometer, the capacity in kW capable of handling the largest tides is

$$N_l = 311 \, AR^2 \tag{3.4}$$

The maximum discharge through the barrage, in m$^3$/sec, is

$$Q_{max} = 57 \, AR \tag{3.5}$$

The power formula applicable for emptying, filling and pumping is provided by the Gibrat-Lewis-Wickert equation

$$E_e = \int_{Z=0}^{H} \gamma f(z) z \, dz \, (f(z) z\_Mz) \tag{3.6}$$

$$E_c = \gamma H \left( V_f + V_{p_1} + V_{p_2} \right) \tag{3.6.1}$$

The energy provided by the flood tide is

$$E_f = \int_{Z=-B}^{H} \gamma f(z) \cdot (H - Z) dz, \tag{3.6.2}$$

Wherein B given levels, by the ebb tide it is

$$E_{ebb} = \int_{z=0}^{+C} -\gamma f(z)(Z - H) dz \tag{3.6.3}$$

Wherein $V_b V_p$ are respectively the capacity (basin volume) and volume pumped,

$E_c E_e E_f E_p$ are the *work* needed for a complete cycle, to empty the basin, fill the basin, to pump. $p_1$ is pumping to lowering and $p_2$ pumping to filling.

+A −A are normal high water level and normal low water level

H is the tide height

B is a water level

f(z) is the area of the basin as a function of water depth

γ is the unit weight of water

Charlier (1982) provided the calculation to find the net energy production for a complete cycle

$$E_c = \gamma H \int_{z=0}^{H} f(z) dz \qquad (3.7)$$

and the difference between operation of the plant involving or not involving pumping.

The formulas for filling, pumping from a level to level B, have been given a.o. in Charlier, *Tidal Energy*, 1983. The gain between involving and not involving pumping is given in (3.8):

$$E_{gain} = (\gamma H(V_{p1} + V_{p2})) \qquad (3.8)$$

Mean extractable power $\bar{P}$ expressed in W/m² is

$$\bar{P} = \frac{1}{2} \mu \omega K_s K_n V^3 = 0.3 \omega V^3 \qquad (3.9)$$

$$\begin{cases} \mu = 0.25, & \text{extraction efficiency factor} \\ K_s = 0.424, & \text{velocity profile factor} \\ K_n = 0.57, & \text{spring/neap tide factor} \end{cases}$$

with as typical values and ω is the density of the fluid in kg/m³.

## 3.5 Range and Amplitude

The difference between high tide and low tide level—*marnage* to the French—varies considerably with geographical location. So does the importance of tidal currents: they dominate circulation in estuaries, though there is some impact of other currents such as aeolean-genic and fluvial currents. Thence, depending on the prevailing tidal range, coasts are said to have a microtidal regime—when ranges remain below 2 m– are not particularly suitable for energy extraction, and waves are the dominant process; or macrotidal regimes when ranges exceed 4 m, and tides are the dominant process. And of course mesotidal regimes are those with ranges between 2 and 4 m; here the dominant process is not as clearly defined.

Microtidal estuaries have usually narrow openings to the sea and may risk occlusion because of sand spit building by sediments carried and deposited by longshore currents. Macrotidal estuaries may be funnel-shaped and tend to be wide.

The tidal prism is the product of the tidal range and the estuary area, which determines the quantity of water entering and exiting an estuary in the course of a tidal cycle. Larger prisms bring about faster currents and deeper channels. Speeds in excess of 1 m/s are common in tidal inlets and estuaries where high tidal ranges prevail. Fast tidal currents are usual where geomorphologic constrictions are present, regardless of the tidal range's magnitude. Indeed, the entire tidal prism must pass though the aperture in six (semi-diurnal tides) or twelve (diurnal tides) hours.

Apogean tidal currents are characterized by a less than usual speed at apogean tides. Those tides, which have a decreased range, occur when the moon nears its apogee. Apogean ranges occur, usually, one to three days after apogee,[2] The tide apogean range is the mean semi-diurnal range occurring at apogean tides time, most conveniently calculated from the harmonic constants Smaller than the mean range in semidiurnal and mixed tide regimes. Where tides are of the diurnal type, the apogean range has no practical importance.

Tide race is used to designate a tidal current (tidal stream) that flows very rapidly where the passage is constricted, *viz.* in a narrow channel. A tide rip is a turbulent body of water resulting from the opposition of a water body to tidal currents (tidal streams); in reality it is a misnomer and the terminology to be used is rip current, even though the current is of small duration and its flow is seaward. It is the return movement of water accumulated near shore by waves and wind.

The tidal phenomenon has been subjected to harmonic analysis and it has in turn allowed identification of "tidal harmonic components (or species)". The major components and their identifying symbols are grouped in semi-diurnal, diurnal and long-period components. The first group components have periods hovering around 12 h; they include the principal lunar ($M_2$), principal solar ($S_2$), larger lunar elliptic ($N_2$), and lunisolar elliptic($K_2$). The diurnal group, whose components have periods of about 24 h, includes the lunisolar diurnal ($K_1$), principal lunar diurnal ($O_1$) and principal solar diurnal ($P_1$). The most influential long-period component is the lunar fortnightly ($M_1$).

The $M_2$, $S_2$ and $M_1$ components are those that most strongly influence spring and neap tides. $K_2$ and $K_1$ result respectively from moon and sun, and earth respective declination. N represents the effect of the moon's elliptical path.

## 3.6 Transmission and Storage

A major problem with the utilization of tidal energy has been the difference between the occurrence of tides and the need for power. It has been a main argument of the opponents and skeptics. Considerable effort has been spent trying to solve the problem and several schemes have been proposed; these include the use of multiple basins in a scheme, compressed air systems, and more. Another reservation has

---

[2] The 1–3 days lag is the parallax age, also known as the age of parallax inequality.

been the problem of power transmission, but the development of high voltage lines has pretty well resolved it, as for instance in Brittany where the Rance plant has been hooked up into the French national grid. Transmission has become, in some cases, less expensive than coal transportation, making ocean-produced electricity competitive. For a 230 kW line as a unit of investment and transport capacity, a 500 kW line will cost 1.66 times as much but will provide six times the transport capacity, and a 8,700 kW costing 2 1/2 times as much will have a capacity 13 1/2 times greater.

Cables of 700 kW can carry overland current for a 2,000 km distance. Capital investment costs drop with increased utilization, but exploitation expenses increase. Transmission losses decrease as tension increases and capacity increases several times faster than tension: 1,000 MW for 400 kV, but 16,000 MW for 1,500 kV. These figures are for alternating current; direct current transmission–such as between the north and south islands of New Zealand–is more economical if the distance involved exceeds 1,000 km. Thought has been given to using cryocables, superconductors and vapor isolation.

Producing electricity for use at a later time can be realized by pumped storage; it is also conceivable by storing compressed gas in exhausted mineral deposit locations, also in aquifers' anticlines. Any gas can be stored in abandoned mines and even artificial cavities can be created for storage purposes. Gorlov's (1982) scheme can be recalled in this regard. Hydrogen can be injected into salt deposits and stored at savings of 80, even 90%, over the cost of using surface positioned tanks. Use of hydrogen as an "intermediary" fuel is, however, not unanimously endorsed (Haydock and Warnock 1982).

Finally, re-timing of tidal energy is possible with electrolytically produced hydrogen to be used as fuel when tide power and peak load demand differ. Hydrogen combustion can heat the air during peaks, and hydrogen can be produced during the off-periods using non-heated compressed air; the hydrogen does neither have to be stored nor transported.

## 3.7 Tides and Harmonic Analysis

Harmonic analysis deals with time series. The repetition of an event—or events—in a cyclic series, is represented by a periodic function which, by the addition of a random variable, become stochastic[3] Tides are cyclic, and oscillatory. A sequence of events is oscillatory when described by a probability "process" and if the trend-free residual element is not randomly distributed.

A smoothing process involving the moving averages method, known as autocorrelation allows estimation of the trend component. An auto-regressive series contains a random component, and the elements of the series are linearly related to the

---

[3] A stochastic function is a probability function containing a random variable.

3.7 Tides and Harmonic Analysis                                                      73

preceding elements of the series by constants. It is impossible to extract from such system oscillations caused by auto-regression once incorporated.

Trend analysis can be performed by obtaining the non-stochastic, harmonic series (which are periodic or cyclic in nature) from a harmonic process containing a random variable. The solution of the harmonic functions, or amplitudes, can be obtained by a Fourier analysis. To get the variance represented in the cycles, a power spectrum analysis is a useful tool.

Perhaps clarification of some of the statistical terms used above might be useful. They are listed hereafter in the order that they appear in the text above.

## 3.7.1 Smoothing

Observations characterized by irregular finite differences, being transformed into a regular succession of finite differences, undergo the process of smoothing. Some observations are thereby lost at the beginning and end of a smoothed succession. It may be compared to discarding "aberrants" at both ends of a statistical population (series). Smoothing tends to hide and shift highs and lows—which may be important because the random component tends to obscure the periodic elements. Though smoothing may artificially generate oscillation in the residual, it is useful for detecting trends and cycles, and can be predictive as with tides.

## 3.7.2 Auto-Correlation

Also called serial correlation, autocorrelation is present, to some extent, in some time series. In correlation, changes in the dependent variable depend on changes in the independent variable; *a contrario* in auto-correlation *changes* in the dependent variable are not accounted for by the independent variable. Successive points in a time series may not be independent from one another, and auto-correlation introduces difficulties—not to be ignored—in time series analysis. That may not be passed over.

## 3.7.3 Moving Averages

Taking a point in a series and adding a number of points on both sides thereof allows to calculate a moving average. In a five point moving total, to the value of the base point are added those of the two adjacent points on either side. Further calculations involve proceeding to the next adjacent point to the base point, and repeating the process. The moving average is obtained by dividing by the total number of points each moving total.

**Table 3.1** Characteristics of various sites appropriate for tidal power plants construction (exclusive of the Rance River TPP)

| | Mean tidal range *meters* | Basin area $km^2$ | Approximate installed capacity *megawatts* | Annual output *TWh* | Plant factor |
|---|---|---|---|---|---|
| **Argentina** | | | | | |
| San José Guff | 5.9 | – | 6,800 | 20.0 | – |
| **Australia** | | | | | |
| Secure Bay 1 | 10.9 | – | – | 2.4 | – |
| Secure Bay 2 | 10.9 | – | – | 5.4 | – |
| **Canada** | | | | | |
| Cobequid | 12.4 | 240 | 5,338 | 14.0 | 0.30 |
| Cumberland | 10.9 | 90 | 1,400 | 3.4 | 0.28 |
| Shepody | 10.0 | 115 | 1,800 | 4.8 | 0.30 |
| **India** | | | | | |
| Gulf of Kutch | 5.0 | 170 | 900 | 1.7 | 0.22 |
| Gulf of Cambay | 7.0 | 1,970 | 7,000 | 15.0 | 0.24 |
| **Korea** | | | | | |
| Garolim | 4.8 | 100 | 480 | 0.53 | 0.13 |
| Cheonsu | 4.5 | – | – | 1.2 | – |
| **Mexico** | | | | | |
| Rio Colorado | 6–7 | – | – | 5.4 | – |
| Tiburón Island | – | – | – | – | – |
| **United Kingdom** | | | | | |
| Severn | 7.0 | 520 | 8,640 | 17.0 | 0.22 |
| Mersey | 6.5 | 61 | 700 | 1.5 | 0.24 |
| Wyre | 6.0 | 5.8 | 47 | 0.09 | 0.22 |
| Conwy | 5.2 | 5.5 | 33 | 0.06 | 0.21 |
| **United States** | | | | | |
| Passamaquoddy | 5.5 | – | – | – | – |
| Knik Arm | 7.5 | – | 2,900 | 7.4 | 0.29 |
| Turnagain Arm | 7.5 | – | 6,500 | 16.6 | 0.29 |
| **Russia** | | | | | |
| Mezen | 9.1 | 2,300 | 15,000 | 50.0 | 0.38 |
| Tugur | – | – | 10,000 | 27.0 | 0.31 |
| Penzhina | – | – | 20,000 | 79.0 | 0.45 |

## *3.7.4 Auto-Regression*

The oscillating residual term results from subtracting the trend from the original data. It is called upon in an analysis of non-periodic terms. Uses include oscillatory phenomena and to identify secondary oscillations in a harmonic phenomenon.

3.7 Tides and Harmonic Analysis

## 3.7.5 Fourier Analysis

Fourier analysis is a method to reduce a cyclic series into its components and then describing the complete cycle by sine or cosine waves. The number of cycles determines the order of the series.

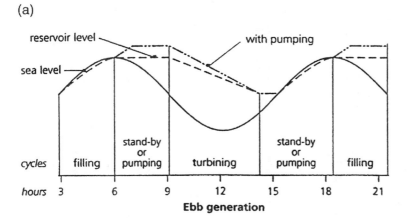

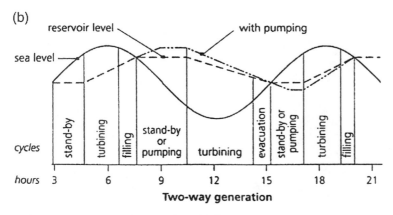

**Fig. 3.1** Alternative operational modes at La Rance, France

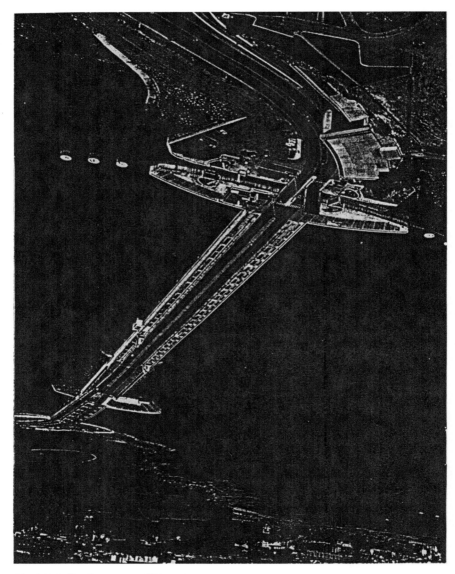

**Fig. 3.2(a)** The Rance River TPP, aerial view

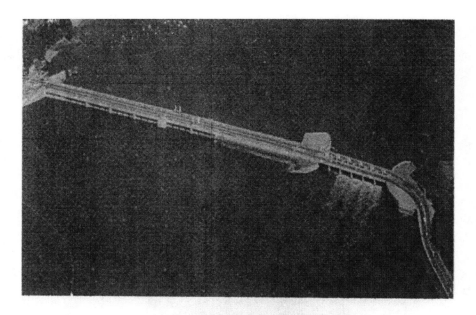

**Fig. 3.2(b)** View of barrage, lock, roadway

# Chapter 4
# Dreams and Realities

Sites are numerous around the world where tidal power could be harnessed. Many have been the locale of tide mills. More have become potential locations as a consequence of the development of low) and ultra-low head turbines. Thorough studies have been conducted in prime areas: Argentina, Australia, Korea, Japan, India, England, Wales. Except for Wales, plans have been laid to rest: Argentina is short of the needed funds, Australia lacks sufficient customers, Korea locked into a political *furore* with France, the potential builder; however, it is again considering building a TPP, this time as an wholly "national" undertaking (Table 4.1).

## 4.1 Dreams?

An old saying has it than often many are called but few are retained. Many plans were made throughout time to build TPP-s, but few were retained....

### 4.1.1 The Severn River and Other British Plants

The Severn River is probably the tidal river which has been most thoroughly studied, and the most often as well, with an eye on building a TPP. It has also been shown that had a plant been built after WW II, it would have quite long ago paid for itself. Because of environmental effects of a barrage, and no doubt the economic aspects of any project, no TPP has ever been even started. Fry (2005)[1] has described alternatives to a TPP-with-barrage. Among these tidal lagoons schemes have been proposed and so has the modest utilization of the tidal stream (tidal current).

Seldom mentioned is that the first utilization of tidal power in the United Kingdom dates back three quarters of a century with a scheme implanted near Bristol (1931), just a few years after the French aborted attempt in Britanny (Aber W'rach).

---

[1] References are to the General Bibliography, "Annex I".

**Table 4.1** Non-comprehensive list of tidal power generation devices

| Device | Characteristics |
|---|---|
| Tide mill | Ancestral scheme including a retaining basin, sluices, wheel, occasionally Sea mill using run of river in an estuary, mostly up and down movement due to tide Most derelict or dismantled. Some reconstructed and resumed activity as tourist attraction or as part of a "working museum". |
| Tidal Power Plant[TPP] | Single or multiple basins system. Single or double (ebb and flood) effect. With or without pumping mode. Requires a barrage or dam. A model with "removable" dam has been designed. Uses bulb, straflo, or even other turbines. |
| Tidal stream Tide current | Uses the tide current (horizontal movement) instead of up-and-down tidal movement. Avoid the need of a dam. Some operate in shallow waters, other schemes in deep fast moving channels. |
| Pulse stream | Developers BMT and IT. Converter uses pair of hydrofoils oscillating across tidal flow, permitting extraction of tidal energy from shallow water. Currently building 100 kW prototype. |
| Gorlov | Aleksandr Gorlov (Northeastern University, Boston) has developed a tidal power scheme with a helicoidal turbine. Needs no dam. Support from US Dept of Energy. |
| E.ON & 8 MW | project on UK west coast. Uses Rotech Tidal Turbines Lunar Energy horizontal-axis turbines mounted offshore on sea-bed. Tide current system to be tested in 1 MW version at European Marine Energy Center (Orkney). |
|  | [EMEC]. Minimal environmental impact (Robert Gordon University). Time frame: 2008–2010. |
| Neptune | underwater generator converting tide current energy for feeding into the grid. A central tower is anchored to seabed. it has 2 arms each supporting a 3-blade rotor. Time frame 2007–2008. |
| Open hydro | US company associated with Alderney Renewable Energy Holdings. 250 kW device to be tested at EMEC. |
| SeaFlow | Property of Marine Current Turbines. Single rotor 300 kW device tried out (2003) off Devonshire coast. Forerunner of SeaGen. |
| SeaGen | 2 rotor device generating 3 times as much power as SeaFlow. (1 MW) Cooperative undertaking EDF, UK Gov., Marine Current Turbines. Time frame 2007–2008. Site: No. Ireland. |
| Lynmouth SeaGen Arrays | 10 MW tidal farm in Bristol Channel. An array of 12 SeaGen-s. Environmental impacts currently under scrutiny. Considered currently at research stage only |
| Tidel | Twin turbo device floating in the tide current, able to turn as the tide itself turns, enabling it to face the flow. Needs no support structure and offering easy installation and maintenance as it is simply moored to the seafloor. Time frame: 2008-2010 for a 1 MW for a pilot trial. Current trials held at NaRec (New and Renewable Research Centre, Newcastle) |

Other rivers have also been considered such as the Mersey. Strong pleas continue to be made, particularly in Great Britain, for harnessing tidal power and the government has implemented various incentive plans towards that end. Details about British past plans can be found in works by Shaw and in Charlier (1982) mentioned in the General Bibliography.

## 4.1.2 Japan

Japan once considered a Kyusyu-sited plant, and currently studies are being conducted at the Tokyo Institute of Technology for sites in Ariake Bay and tidal current schemes in the Kuroshyo. Tides in Ariake Bay reach 4.56 m with a mean range of 3.18 m. Mean depth of water is 21 m and maximum depth is 40 m. A barrage built to span the bay would be 13.6 km long. The output with 100 generators (compared with the Rance's 24) would be 500 MW. Annual electricity generated would be $68.4 \times 10^8$ kWh. The price tag, in the seventies, for such a plant was estimated to exceed $2,5 \times 10^{10}$ Yen, around US$2,500 million.

## 4.1.3 South Asia, Egypt

Studies have been pursued in Bangladesh (Salequzzaman and Newman 2001) while Singapore examined four locations to tap the energies of tides: the Lighthouse, Pasir Panjang, Sembawang, and Tanjong Fagar (Gupta et al. 2001). A design for a TPP in Egypt on the Red Sea has been submitted by Fahmy (2001) at the Fifth International Conference on Electrical Machines and Systems held in Shenyang (PR China) in 2001. As in China, the locations for the proposed plants are remote, hence these would provide small quantities of power, ranging from 1 to 10 MW.

## 4.1.4 Down Under

What has occasionally been hailed as the largest electrical power potential in Australasia remains untapped. One site, some 250 km north of Broome (State of Western Australia), was already considered as a "source of electricity" nearly 90 years ago.[2] It was also discussed in a thesis in 1950 (Raynor) that dealt with tidal power in general. Since then and through the 1970s, the matter periodically received some publicity and the area was the subject of an in-depth study[3]. The latter dates from the mid 1960s and is the subject of SOGREAH reports published both in Perth and in Grenoble (France).[4] Of 50 sites examined along 1,700 km of coast, half were

---

[2] Easton, W.R., 1921, *Report on North Kimberley District:* Perth (West. Austr.), Northwest Department Government.

[3] Lewis, J.G., 1963, The tidal power resources of the Kimberleys: *J. Inst. Eng.Austral.* 35, 12, 333–345; Maunsell et al.,1976, *Kimberley tidal power:* Perth (W.A.), The State Energy Commission of Western Australia; Saunders, D.W., 1974, Kimberley tidal power revisited: *The Inst. Eng. Austral., Conf.* [Proc. Techn. Conf.] Melbourne (Vict.) 47, 11, 44–51. 47–55; Saunders, D.W., 1976, Kimberley tidal power: *Proc. Cong. Austr. New Z. Assoc. Adv. Sci*; Scott, W.E., 1976, Australia takes a new look at tidal energy: *Ener. Int.* 13, 9, 41–43.

[4] SOGREAH=Société Grenobloise d'Applications et d'Etudes Hydrauliques (constructors of the Rance TPP for the French electricity company).

found technically (but obviously not economically) suitable, with Secure Bay, Walcott Inlet, George Water and St George's Basin as the preferred ones.

### 4.1.5 Much Power, No Users

At several Australian sites, tidal ranges exceed that of the Rance, running from 9 to 12 m. TPP locations exist also on the east coast, e.g. on Broad Sound. The feasibility study, while mentioning several big problems—that however "can be overcome"—points out that once a plant is constructed, it would be rather simple to expand the scheme to satisfy additional demand. The SOGREAH recommendation foresaw 30 bulb-turbines, inclusion of islands to cut down cost of the dam, two sluiceways, single generation, and an awesome ten-year construction window. There would be cofferdams—no longer considered necessary, as explained in Maunsell's study—and a kilowatt would still be considerably more expensive than one generated from fossil fuel. However, Lewis considering a rapidly amortized loan (for construction) by a sinking fund, predicted produced power below the cost of that from a nuclear plant.

There have been no developments on the Australian tidal power horizon for some decades. Possibilities were also examined for New Zealand, but these are dwarfed by Australian potential.[5] The problem is that there is plenty of power available, and harnessable, but there is a lack of potential close-by customers in this region of Australia, and is similar to the one in Northwest America where energy is for the taking, but customers are too few; it is what has torpedoed any plans to implant a TPP on the Turnagain Inlet.[6]

### 4.1.6 India

There have been new statements of interest made by Indian laboratories, but no actual plan is being "pushed". Of the several locations once retained, the Gulf of Kutch is the principal one. The 600 MW installed capacity plant would have necessitated three barriers, a 3 km one across Hansthal Creek. The help of the Electricité de France would have been called in. The Rann of Katchchh was ultimately selected for a feasibility study. As in France, proponents envisioned a road atop the barrage that would encompass no less than four barriers. The scheme would be a single-basin, single-effect facility taking advantage of an 8 m local tide range.

But some drawbacks were obvious: the vicinity of the ports of Navlakhi and Kandia, the lack of a solid firm rock foundation to within 15 m of the site with a clayey soil as foundation, aeolean sediments and a highly saturated marine littoral,

---

[5] Keough, Mc., 1959, Tidal power: *NZ Elec. J.* 32, 3, 82–83.
[6] See details, e.g. in Charlier, 1982, *Tidal energy:* New York, Van Nostrand-Reinhold.

conspicuous sedimentation, an uneven topography, seasonal variations.[7] On the positive side substantial socio-economic advantages would accrue and the absence of natural impact has been underscored. Optimists had seen the plant "at work" by 1997, but more than ten years later not even the first stone has been laid.

## 4.2 Realities

There is an expanding literature about China's numerous mini-plants yet very little is actually known about them (Charlier 2001). There is no shortage of announcements coming from China, Korea and Russia about, mostly large, new plants. But these remain *pronunciamentos* at international congresses and in press releases, so far all *sans lendemain*. France shelved its mega-project of the Chaussey Islands, but then the Electricité de France has been on a nuclear binge and even if the company lost its monopoly (2007) under threats of the European Comission's competition rules, nuclear power remains a preferred option of France's energy policy. If environmentalists raise the same fears of disturbed ecosystems, not so-environmentally-benign river barrages are built.

Few large plants became a reality (France, Russia, Canada); many small ones were built with modest means in China. Ephemeral plants are reported to have functioned in a variety of locations. A central in Boston Harbor was apparently short lived because of expansion of the port and today is a prized real estate area. Another one has disappeared in Suriname where it once was located on Lake Van Bemmelen. Near Bremerhaven (Germany), a tidal plant folded when World War I rumblings filled the news.

Construction was once started in Brittany on the Aber Wrac'h but aborted when the economic crisis of the nineteen-thirties ran havoc with many a project, and the Quoddy dreams ran into the US Congress' ears listening to the siren's whistles of the power industry. And so actually only one large TPP was build, and several pilot and mini-plants.

Few engineers' dreams came thus to pass.

### *4.2.1 The Rance River Plant*

Located on the coast of Brittany the largest tidal power station includes a barrage—with fish pathway—that spans the Rance River Estuary at about its narrowest site (750 m). A 14 m two-lane highway has been constructed atop the dam.[8] Its

---

[7] Sharma, H.R. & Vats, T.P., 1985, Environmental impacts of the proposed Katchchh Tidal Power Plant: *Proc. Int. Sem. Env. Assmt, Water Research* at Roorkee, India; Vaidyaraman, P.P. et al., 1986, Experience during field data collection for tidal power project in the Gulf of Katchchh: *Int. Symp. Wave, Tidal, OTEC & small scale HydroEnergy* 3, 15, 191, 201.

[8] Problems due to it seem to have arisen since 2006.

approximate cost is a hundred million 1960-US$. Equipment ran 55% of cost and cofferdams 13%; the latter expenditure has been avoided in the construction of later plants. The semi-diurnal tides reach approximately 13.5–14 m. The plant operates on a combination double-effect and pumping scheme which provides for a production "tailored" to time of day and tidal amplitude, favoring the solar cycle. The maximum output is 240 MW and annual power generated is 544,000 kWh.

Inside the plant 24 bulb groups of 10,000 kW have been installed: bulb groups have reversible blades allowing generation in both ebb and flood directions. Power is transmitted to Aube, Brest, Landerneau, Rennes and Paris. As long as the head remains minimum 7 m the plant has in a basin to sea direction a capacity of 10 MW; it is much lower in the other direction. The purpose of the pumping function is to increase the head near high tide by storing water; near low tide pumping allows to over-empty the reservoir below sea-level. Pumping is not necessary for high tidal ranges (9 or 10 m).

In France obviously the dream came through. It is examined in detail in the next chapter.

## 4.2.2 The Kislaya Bay Plant (Russia)

Kislaya Bay (Kola Peninsula) is a White Sea inlet, near World War II famed Murmansk in the Russian Arctic. It is in the Arctic that most of Russia's tidal energy is dissipated. The dam was built by assembling on the site pre-fabricated sections between 40 m high cliffs. A narrow canal (40 m) establishes the communication with the sea. Materials adapted to the harsh climatic conditions were developed for use in construction (igneous rocks aggregates mix; epoxy resin insulation reinforced with glass-fabric). The depth is merely 3–4 m, a factor that has caused environmental problems (Karasev et al. 1996, Marfenin et al. 1997).

Precast concrete cellular units were built on land, brought to the site and sunk into place by filling some cells with sand. The pilot plant uses the bulb type turbines also installed in the Rance River station.

Tides range here from 9 to 13 m. At the plant the normal maximum is 11 m. The head is 15.2 m. The plant's capacity is 400 kW. The small bays of the country hold a power potential exceeding 3.2 billion kW ($3.2 \times 10^9$ kW). It is far smaller than that on the Rance River measuring only $36 \times 18$ m with a height of 15 m. The site was also selected because of its vicinity to the existing power system.

Total electricity generated is estimated at 1,800 kW. A more detailed analysis of this major plant is part of a further chapter.

Even though 2006 dated information pertaining to tidal energy use in Russia indicates a current disinterest, Usachev et al. published in 2004 both a review of its large-scale use and the observations made over 35 years of operation of the Kislaya Bay TPP. The TPP has operated successfully, with moving parts underwater, under extreme climatic conditions. The plant has been used as a biological test site in a nearly closed-off basin, including fisheries and mariculture observations. A variety

4.2 Realities 85

of units were tested among which cathodic behavior, sorption and electrolytic. The reporters confirmed that a TPP can be integrated into a power system, in basic and peak periods of the load diagram, allowing for the specific conditions of tidal energy generation.

### 4.2.3 Annapolis-Royal Pilot Plant (Canada)

The Annapolis Royal plant, located on Hog's Island in the Bay of Fundy, in the middle of the intake canal supported by a vertical concrete pier on the lower Annapolis River, takes advantage of a 7 m tidal range. The plant's powerhouse is small (25 × 6.5 × 15.5). It was built at a cost of 1982-C$46 million. A cofferdam was used.

A single-effect Straflo turbine was installed in this pilot plant. Such turbine was developed in Switserland in 1974 and is arranged in a horizontal water passage with the (rim type) generator field poles attached to a motor rim mounted around the periphery of the propeller; turbine and generator form a coupled unit without power shaft, saving thereby on powerhouse costs. Production is 50 million kW.

The dam is 225 m long. A fish pathway has been included, but apparently has not curbed mortality in a very significant manner (Dadswell and Rulifson 1994).[9]

### 4.2.4 A Hundred Chinese Plants

According to the relevant Chinese literature 128 plants would be in operation in China with a total capacity of 7,638 kW. The largest plant, a multiple basin complex on the Shunte River, has a capacity of 304 kW. The same sources claim that the 40 kW Zhejiang Province plant predates (1959) the Rance Plant. Maximum heads range from 3.49 to 7.8 m, except for Jian Xia station where the head reaches 8.39 m; that plant and the Liu He Kuo one are double-effect schemes.

A review of the rather limited available information on Chinese plants has been recently published (Charlier 2001, "The view from China") and, in the same issue of *Renewable & Sustainable Energy Reviews,* Chang et al. (2003) review all aspects of the energy scene including tidal and tidal flow energy. The paucity of information about Chinese TPPs has been mentioned. Descriptions have often appeared in rather obscure journals, that, naturally, are frequently Chinese. Of the Chinese plants, the Kianghsia facility, however, was briefly described in the [French] *Revue de l'Énergie.* The first Chinese TPP is, size-wise, far less important than the Rance River TPP (240,000 kW) with six 500 kW turbines that will deliver 10,740 kWh annually. (Song 1981).

Not to be outdone, Taiwan, similarly, examines various sites and systems for modest-sized plants (Hau Ku 2000, Shern et al. 2003) Deviating from the traditional

---

[9] See also chapter on "Tidal power in the Americas", and Annex IV.

tidal power plant, a slightly different scheme was proposed for Taiwan (Jwo-Hwu 1998). It was suggested to generate power from buoyancy and weight at ebb tides. The system apparently can be used anywhere where there is a head, whether a coast, a dam or a pumped-storage reservoir. Perhaps a most significant feature is the ability to shift off-peak generation capacity to the peak need period. The time delivery shifting problem has been one of the contentious topics of tidal power.

### 4.2.4.1 Tide Current

The current keen interest in tapping the tide current has generated in several locations thought of revisiting tidal power particularly where the site is remote and the demand small. The past interest of an American–Indian tribe in the Quoddy region may perhaps be revived, even if the reservation's location is not in a thinly populated area.[10]

---

[10] See Charlier, R.H., 1982, *Tidal energy:* New York, Van Nostrand-Reinhold; Laberge 1978.

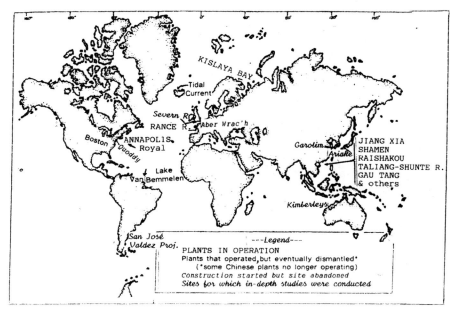

**Fig. 4.1** Location of plants in operation or dismantled, or aborted and sites studied in-depth

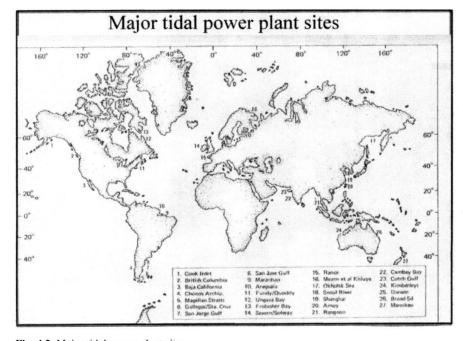

**Fig. 4.2** Major tidal power plant sites

**Fig. 4.3** Work proceeded at Rance River site inside cofferdams

**Fig. 4.4(a)** View of Rance R. TPP. Chalibert Island is in foreground

**Fig. 4.4(b)** View of Rance R. TPP

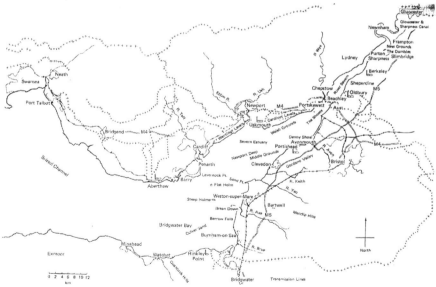

**Fig. 4.5** Location map Severn R. estuary and site proposed TPP

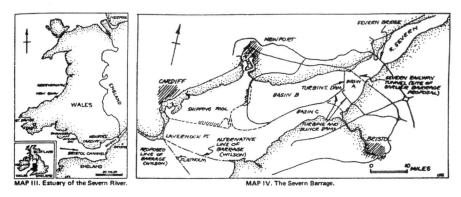

**Fig. 4.6 (a, b)** Detailed maps of proposed Severn R. TPPs

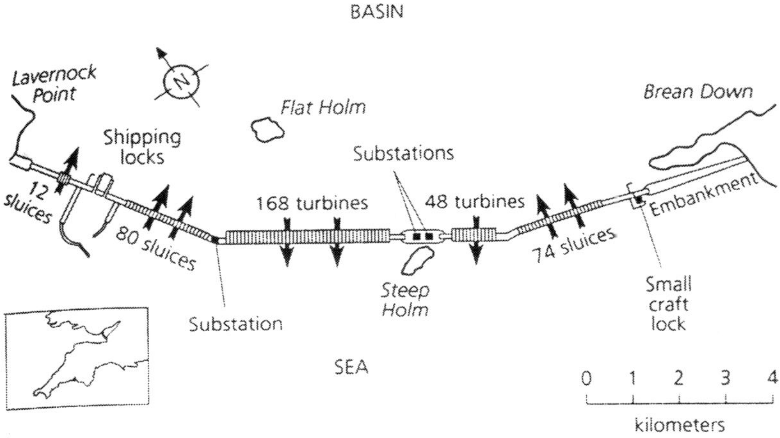

**Fig. 4.7** Proposed TPP scheme for the Severn River (Wales)

**Fig. 4.8** Mock-up of the Kislaya Guba TPP (near Murmansk, Russia) as exhibited in Tokyo by USSR embassy

**Fig. 4.9** View of the Kislaya Bay TPP

**Fig. 4.10** Kislaya Bay, USSR. Artist's view

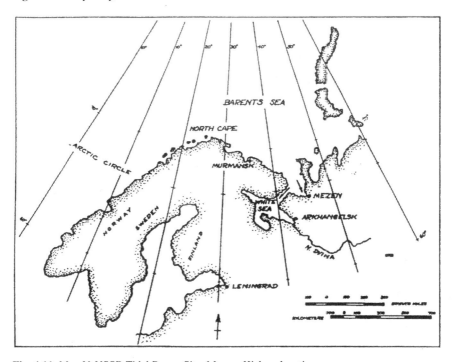

**Fig. 4.11** Map V. USSR Tidal Power Site. Mezen, Kislaya location

**Fig. 4.12** Powerhouse being towed to site

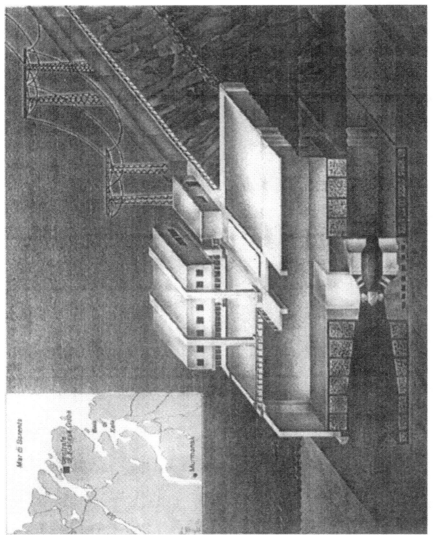

**Fig. 4.13** Location and artist's view of Kislaya TPP

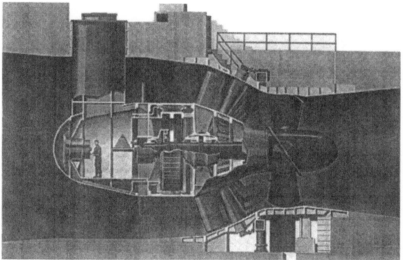

**Fig. 4.14** Bult turbine installed at Kislaya TPP

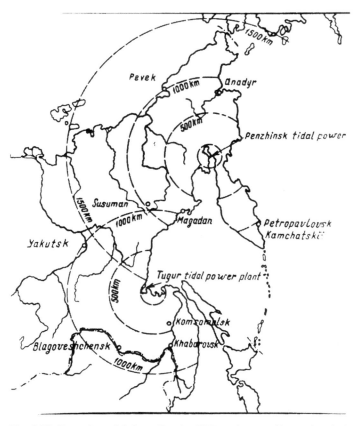

**Fig. 4.15** Sites of possible large Russian TPPs and areas of large electrical consumption

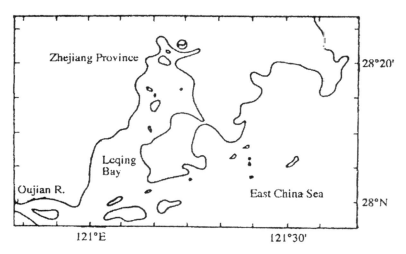

**Fig. 4.16 (a)** China—Location map of tidal power on Leqing Bay in Zhejiang Province

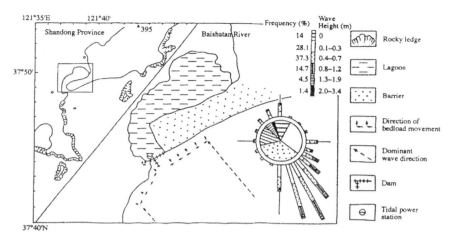

**Fig. 4.16 (b)** The Raishakou tidal power station (P.R. China)

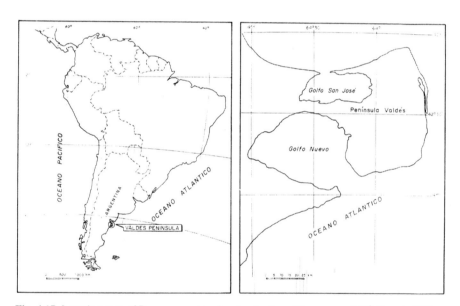

**Fig. 4.17** Location map of Passamaquoddy showing basins of the proposed US-Canada TPP

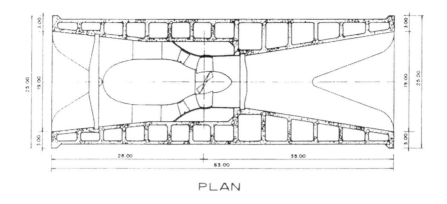

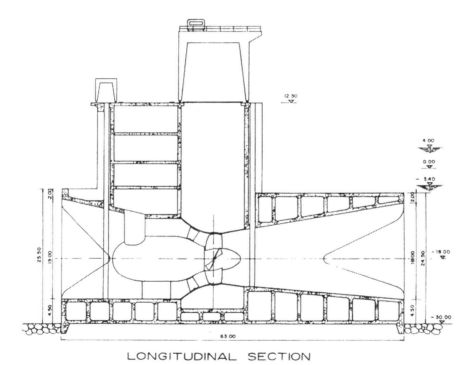

**Fig. 4.18** Bulb unit generating caisson

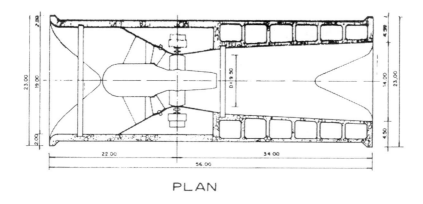

PLAN

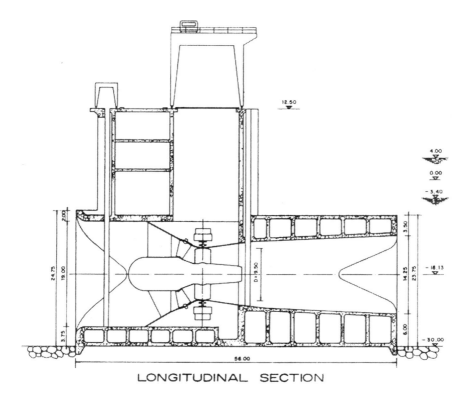

LONGITUDINAL SECTION

STRAFLO GENERATING CAISSON

**Fig. 4.19** Straflo generating caisson

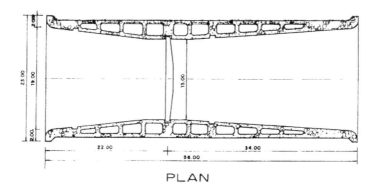

PLAN

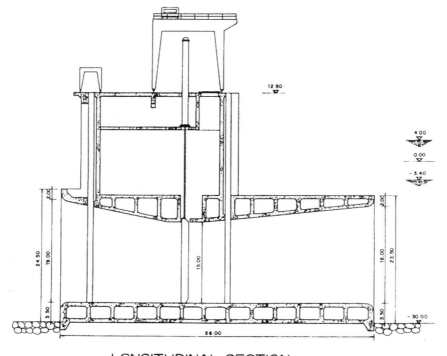

LONGITUDINAL SECTION
SLUICEWAY CAISSON

**Fig. 4.20** Sluiceway caisson

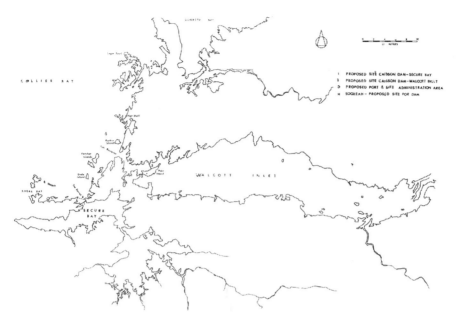

**Fig. 4.21** Sites map. Australia. Kimberley Region TPP project

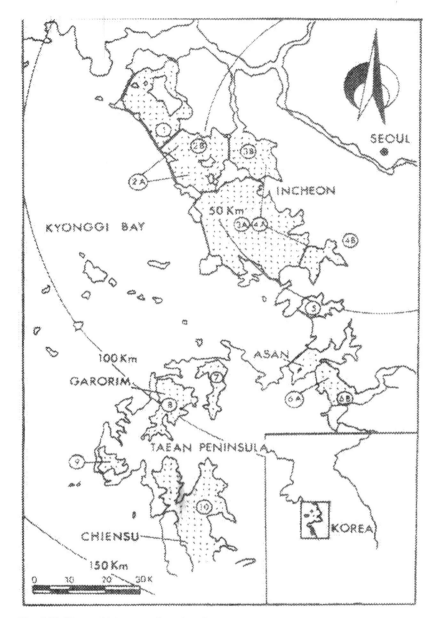

**Fig. 4.22** Korea: tidal power plants location map

# Chapter 5
# The Anatomy of the Rance River TPP

## 5.1 Introduction

The price rise of traditional fuels, particularly oil and gas, and the concern for global warming, have been strong incentives for a renewed look at alternative sources of energy. Though only a modest potential contributor to alleviation of the "crisis", attention has nevertheless been again directed towards ocean generated power. A major anniversary was celebrated in 2006 for the largest tidal power plant currently in existence.

Publications dealing with turning the tidal energy into power spread over the better part of the 20th century; we could venture to say, without erring, even *ad libitum*. Among the more recent ones focusing on detailed edxamination of the basics, the principles and the economics are the contributions of Baker, Charlier and Gibrat.[1]

The economics have changed since the 1960s and 1980s books by these authors.

Many other authors, of course, significantly added to the basic knowledge of tides and tidal power[2].

Even some exquisite blueprints are to be found in the rather recently translated and published *Notebooks* of Leonardo da Vinci, and in the writings of Mariano[3]. The literature addressing specifically the Rance River tidal power plant is far less abundant, but far from negligible[4].

---

[1] Baker, G.C., Tidal power. Some historical implications. *Proc.Conf. New Approaches to tidal power. Bedford Inst.Ocean.* 1982, I, 1–6; Baker, G.C., *Tidal power.* London, Plenum Press 1982; Charlier, R.H., *Tidal energy.* New York, Van Nostrand-Rheinhold; Charlier, R.H. and Justus, J.R., Is tidal power coming of age? **In** Charlier, R.H. & Justus, J.R., *Ocean energies.* Amsterdam, Elsevier 1993 [Chap. 7]; Charlier, R.H., Sustainable co-generation from the tides. *Ren.Sust.Energy Rev.* 2003, **7**, 3, 187–213; Gibrat, R., *Les usines marémotrices.* [in French: tidal power plants] Paris, Electricité de France 1955.

[2] Charlier, R.H., Sustainable co-generation from the tides. *Ren.Sust. Energy Rev. A bibliography,* 2003, **7**, 3, 215–247.

[3] Richter, J.P., *The notebooks of Leonardo da Vinci.* New York, Dover 1970 [English translation]; Mariano, *Potentiae ædus usus* [in Latin: Utilization of the power of the tide].Siena (anno 1438).

[4] E.g. Gibrat, R., The first tidal power station in the world under construction by French industry on the Rance River. *French Technical Bulletin* 1962, **2**, 1–11; Gougenheim, A., The Rance tidal energy installation. *J. Inst.Navigation 1967,* **XX**, 3, 229–236; Jones, J.E., The Rance tidal power station. *Geography* **53**, 11, 412–415; Kammerlocher, L., La station marémotrice expérimentale de

## 5.2 Ancestors and Forerunners

Tapping the energy of the ocean's tide and of the river's tidal current to produce mechanical power is a practice that goes back centuries[5]. A conventional tide mill included sluices, a dam, and retaining basin if it was not a run-of-the-river type mill, and wheels proved to be quite sophisticated equipment for the times. Often a mill included several buildings such as the house of the miller. Though dreamt about for more than a century, only forty years ago was the first "sizeable" plant to capture the energy and turn it into electrical power built and placed in service. Since then small plants have been providing power in Russia, Canada, and China.[6] The French plant has provided reliable, sustainable, highly useful power, besides several other regional and national advantages. Every ten years or so, an analysis of its performance has been published, the last one in 1997.[7]

The performance, improvement, updating of the Rance River facility has been consistently reported: Barreau and Charlier described it when the plant celebrated 30 years of existence, Booda (1985) after respectively 20, 15, 10, and 6 years of functioning. M. Gandon, discussed also 6 years of operation (1973), so did Wilson (1973) and Gandon et al. (1973). Corrosion had been addressed by Faral (1973), Leborgne (1973), and Duhoux (1973), cathodic protection by Legrand (1973).

Putting tides to work for power production has been proposed through four major methods. The most common idea consists in the float method whereby the incoming tide would raise a floating mass that, as it falls back to its original position could do useful "work". In a second approach a rotating paddle wheel is mounted on a shaft and activated by both ebb and flood tides, with power transmitted by a shaft.

---

St Malo. [in French: the experimental tidal power station of St Malo] *Rev.Gén. de l'Electr.* 1960, **69**, 5, 237–261; Mauboussin, G., Construction de l'usine marémotrice de la Rance. Contribution des essais sur modèle réduit à la mise au point d'un mode d'exécution des travaux. [in French: Rance tidal power station construction. Contribution of trials on reduced scale model of the work execution mode]. *La Houille Blanche-Rev.Int. de l'Eau (4es Journées de l'Hydraulique)* 1957, 388–399.

[5] Charlier, R.H. & Ménanteau, L., The saga of tide mills. *Ren. Sust. En. Rev.*, 1997, **1**, 3, 271–207; Charlier, R.H., Chaineux, M-C.P.& Ménanteau, L., Rise and fall of the tide mill. In Morcos, S., Zhu, M., Charlier, R.H. et al. (eds) *Ocean sciences bridging the Millennium.* Paris, UNESCO & Qingdao, Ocean Press 2004 pp. 314–338.

[6] Bautier, A.-M., Les plus anciennes mentions des anciens moulins hydrauliques et de moulins à vent. *Bull. Philolog. & Hist.* 1960, **2**, 590–592 [in French: the oldest mentions of the old hydraulic mills and of wind mills]; Charlier, R.H., *Tidal energy.* New York, London, Melbourne, Van Nostrand-Rheinhold, 1982; Charlier, R.H. & Justus, J.R., *Ocean Energies: Environmental, Economic and Technological Aspects of Alternative Power Sources.* Amsterdam, London, New York, Tokyo, Elsevier Science Publ., 1993 [Oceanography Series]; Charlier, R.H., Ocean alternative energy. The view from China: small is beautiful. *Ren.Sust.Energy Rev.* 2001, **3**, 3, 7–15; Ch'in Hsu-Ts'ung, The building of the Shamen TPP. *Tien Chi-Ju Tung- Hsin* 1958, **9**, 52–56; Zu-Tian, G., The development of tidal resources in China: the tidal power experimental station of Jiangxca and its Nr 1 and Nr 2 bi-directional tidal water turbine. In Krock, H.J. (ed.) *"Ocean energy recovery" Proc. Int. Conf. ICOER* 1989. Amer. Soc. Civ. Eng., Ocean Energy Div. Pacific Int. Center for High Technology Research, University of Hawaii-Manoa pp. 157–166.

[7] See fn 13, Barreau.

The third system is already more sophisticated; its use has been suggested in contemporary schemes, a.o. by A.M. Gorlov of Northeastern University (Boston)[8]. Air contained in a conduit of metal or concrete is compressed by the incoming tide. The compressed air can be called upon at any time and the system thus frees the plant from the constraints of the lunar cycle. Gorlov has improved on his approach and publicized it in a seminar held at and organized by Northeastern Uuniversity in 2005. And a fourth approach involves damming in part of the sea, providing a basin that fills up at incoming tide; the water is then released at low tide, forced to pass through turbines either back to the sea or, in multi-basins schemes, to another basin.

With oil prices having leaped up to levels surpassing $120 a barrel, the interest for ocean energies has rekindled. The cost of an ocean-generated kilowatt has become attractive. Even some thought is given to revive the tide mill, already a touristic attraction since several decades, and blossoming.[9]

Tidal power stations could provide valuable co-generation, yet only very modest stations have been built since the Rance River station in Russia, Canada and China[10].

Much attention is now given to tapping tidal current energy, which dispenses with a barrage, a system even closer to the ancient run-of-the-river tidal mill[11]. Talk about putting tide mills, ancient, rebuilt or new, back to work, may not be entirely idle, as these "traditional facilities" could perhaps, in isolated sites, refloat decaying villages' industries.

## 5.3 Tide Mills Bow Out on the Rance

Half a century ago tide mills had not entirely disappeared from the power-producing scene. Some isolated ones were still in operation in the British Isles, on the Iberic Peninsula and in the French province of Brittany. Several worked, coincidentally, on the Rance River estuary. The abers of Brittany had always been a privileged site for such water mills and the Aber Wrac'h had been the site of the ill-fated first attempt at constructing a tidal power plant in France.

The "remnants of the past" were dismantled in the sixties to make room for a rather large electricity power plant generated by ocean energy. Their removal may, however, not be the last hurrah and there might perhaps be a modest renascence in the not so distant future.

---

[8] Gorlov, A.M., 1979 Some new conceptions in the approach to harnessing tidal energy. *Proc. 2nd Miami Int. Conf. On Altern. Energy Sources*, pp. 1171–1195.

[9] Tuttel, J., Watermolens, eeuwig oud en eeuwigbloeiend. *Aard en Kosmos*, 1978, **10**, 8–9 [in Dutch: water mills, eons old and ever blossoming].

[10] Charlier, R.H., 2003, Co-generation from the tides. A review: *Renewable & Sustainable Energy Reviews* **7**, 3, 187–213.

[11] Charlier, R. H., 2004, A sleeper awakes: power from tidal currents: *Renew. & Sustain. Energy Rev.* **8**.

## 5.3.1 The Rance River Plant

The Rance River tidal power station transposed the tide mill principles of tidal energy harnessing into modern times. A retaining basin or a bay may be closed off from the river by a barrage or dam; in tidal stream or tide current schemes no barrage is involved, a part of the river flow is diverted and forced to pass through a separate channel.

At the Rance River station both ebb and flood currents can generate electricity and even some power is generated by pumping. The nine to fourteen meters range tides can produce close to 550,000 kW. The station is linked to France's National Electricity Grid and has played a major beneficial economic, and social, role. A highway crosses the river on top of the dam, which shortened by more than 30 km the distance between towns on both sides of the Rance. Some problems have been reported in 2006 with the road.

### 5.3.1.1 Principles Involved and Choices to Be Made

The site selected for the first large sized tidal power station is one where several tide mills had functioned for centuries and some were still at work when the project got the green light.[12] Designed in 1959, the Rance River power station went into operation seven years later. Construction costs ran some $100 million (€ 145 million at 2003 exchange rate) but opened up a rather impoverished, somewhat backward, Brittany and allowed it to "pull itself up into the 20th century"; concomitantly it tolled the demise of the last tide mills still operating on the Rance.

Three precise sites were eventually retained to locate a tidal power station. The system chosen by the Rance River Plant builders[13] is only one of four "basic" ones. The first one is that of a basin filling at flood tide with power produced at ebb tide only. The second uses multiple basins and simple turbines make continuous generation possible. The third approach is that of a single basin but by the use of pump-turbines ebb-and-flood generation is possible, power can be provided on demand and independently of the tide state. Finally, with a single basin, simple turbines, a high head pumped storage arrangement permits continuous, not-tide-connected, maximum efficiency production. The latter is the most expensive scheme to operate.[14]

---

[12] Whether the Rance River plant was the first is sometimes challenged, but it certainly was the first of large dimension and of high production. A tidal power station functioned in the latter part of the 19th century in Boston (Massachusetts, USA) harbor [cf. H. Creek, 1952, Tide mill near Boston: *Civil Engineering* **22**, 840–841]. Another was at work near Husum, Germany but was dismantled when World War I started (1914). Some Chinese authors have claimed that a tidal power station existed in China prior to the construction of the Rance River station [Ch'in Hsu-Ts'ung, 1958, The building of the Shamen TPP: *Tien Chi-Ju Tung-Hsin* **9**, 52–56].

[13] SOGREAH=Société Grenobloise d'Applications Hydrauliques.

[14] Wickert, G., Tidal power. *Water Power*, 1956, **8**, 6, 221–225; Wickert, G., Tidal power. *Water Power*, **8**, 7, 259–263; Wilson, E.M., Energy from the tides. *Science Journal*, 1965, **1**, 5, 50–56.

## 5.3 Tide Mills Bow Out on the Rance

The Rance River station (Brittany, France) has a retaining basin, a barrage or dam and sluices. Twenty-four bulb type turbines are placed in the dam.[15] Plant turbines can be of the *bulb* or *Straflo* rim type and even other designs have been tested. A large head was originally required for a plant to properly function but technological advances permit to use low-head turbines nowadays. Both ebb- and flood-currents can generate electricity, an achievement that was at the time heralded as remarkable, but whose extra cost is unjustified according to current thinking by some engineers.[16]

The 9–14 m range tides can produce close to 550,000 kW. The station is linked to France's national electricity grid. It opened up a rather impoverished Brittany region and allowed it to pull itself up into the 20th century, provided needed power to France as well, and the four-lane highway built on top of the barrage has shortened travel between the two banks of the Rance by several hours. The site, as mentioned previously, is one where several tide mills had functioned for centuries. Its construction tolled the demise of the last tide mills on the Rance. Designed in 1959, the electricity station went into operation seven years later. Construction cost $100 million (approximately €66 million at 2008 exchange rate).

The French station nears its half-century anniversary. It performed well, caused minor environmental impacts, lifted an entire region out of an enduring

---

[15] The Rance and Russian plants use bulb turbines, the Canadians installed Straflo types. *Cf.* Berns[h]tein, L.B., Kislogubsk: a small station generating great expectations. *Water Power,* 1974, **26**, 5, 172–177; Charlier, R.H., *Tidal Energy.* New York, Van Nostrand-Reinhold, 1982, pp. 114–119; De Lory, R.P., Prototype tidal power plant achieves 99% availability. *Sulzer Techn. Rev.,* 1989, **1**, 3–8; De Lory, R.P., The Annapolis tidal generating station. *3rd Int. Symp. Wave, Tidal, OTEC and small scale hydro energy,* 1986, **III**, 125–132; Douma, A. & Stewart, G.D., Annapolis Straflo turbine will demonstrate Bay of Fundy tidal power concept. *Modern Power Systems,* 1981, **1**, 53–57; Gibrat, R., L'usine marémotrice de la Rance. *Rev. Franç. de l'Energie, 1965 (Num. d'avril)* [in French: The Rance tidal power station].

[16] André, H., Ten years of experience at the "La Rance" tidal power plant. *Ocean Manag.,* 1978, **4**, 2–4, 165–178; Banal, M. and Bichon, A., Tidal energy in France—The Rance River tidal power station—Some results after 15 years of operation. *J. Inst. Energy,* 1982, **55**, 423, 86–91; Barreau, M., 30th anniversary of the Rance tidal power station: *La Houille Blanche-Rev. Int. de l'Eau* 1997, **52**, 3, 13; Bonnefille, R. & Thielheim, K.O. Tidal power stations. 1982 **in** *Primary energy. Present status and future perspectives,* Heidelberg, Springer Verlag pp. 319–334); Booda, L.L., River Rance tidal-power plant nears 20 years in operation. *Sea Techn.* 1985, **26**, 9, 22; Charlier, R.H., French power from the English channel. *Habitat,* 1970, **XIII**, 4, 32–33; Charlier, R.H.,Comments to "Power from the tides" by C. Lebarbier. *Nav. Eng. J.,* 1975, **87**, 3, 58–59; Charlier, R.H., Re-invention or aggiornamento? Tidal power at 30 years. *Ren. Sust. Energy Rev.*1997, **1**, 4, 271–289; Cotillon, J., La Rance: six years of operating a tidal power plant in France. *Water Power, 1974,* **26**, 10, 314–322; Wilson, E., Energy from the sea—Tidal power. *Underwater J. & Infor. Bull.* 1973, **5**, 4, 175–186; Hillairet, F., Vingt ans après. La Rance. Une expérience marémotrice. *La Houille Blanche-Rev. Intern. de l'Eau,* 1984, **8**, 571–582 [in French: Twenty years later. The Rance. A tidal power experience]; Hillairet P. & Weisrock, G., Concrete benefits from operational tidal power station. *Int. Symp. Wave, Tidal, OTEC and small scale hydro energy*[Brighton, UK],1986, **III**, paper nbr 13, 165–177; Lebarbier, C.H., Power from tides: the Rance tidal power station. *Nav. Eng. J.,* 1975, **83**, 2, 57–71; Lebarbier, C.H.,Power from the tides. Discussion. *Nav. Eng. J.,* 1975, **87**, 3, 52–56; Retiere, C., Bonnot-Courtois, C., LeMao, F. & Desroy, N., Ecological status of the Rance River Basin afterv 30 years of operation of a tidal power plant. *La Houille Blanche. Rev. Int. de l'Eau,* 1997, **52**, 3, 106–107.

slumber, and delivered needed extra power to distant areas. Many lessons have been learned covering both construction and operation. True, the stations built since 1966 have been experimental and/or pilot plants; a major deterrent has been the capital-intensive aspect of such undertakings and the resulting high cost of the delivered kilowatt. However, the cost reducing approaches have lowered the price of power production, further reduced when taking into consideration the life cycle of the plant.

## 5.3.2 Other Anniversaries

Other birthdays are to be or even have, been celebrated. Skipping plants which have functioned before World War I, in Massachusetts and Germany, if we are to give credence to some Chinese claims, a tidal power station was built and put in service already in 1959. That would make it 47 years old. The Russian plant near Murmansk reached 40 in 2008 and the Nova Scotia facility celebrated recently its 30th anniversary.[17]

Tidal power stations, such as the Rance, can provide valuable co-generation but the other existing stations are all very modest[18]. If proponents of tidal stations have shied away from gigantic schemes, such as the Chausey Islands Project[19] for instance, it has not stopped engineers to dream of huge schemes, some proposing to construct a granite barrage across the English Channel. Closer to the present is even a major scheme put recently on track in Korea, where different locations have been repeatedly under consideration.

No less than 21 sites had been under consideration, in France, the earliest one being the Bay of Arcachon (1902), but which would have perhaps wrought havoc for the flourishing oyster producing industry. All the early sites were on the Channel, the Dover Straits and the Atlantic. Strangely, up to the very last moment a larger plant had been favored near the Mont St Michel on the border of Normandy, location of a splendid abbey, and a major tourist attraction. The latter may have played a role in the ultimate choice.

---

[17] See fn 6, 12 and 13.

[18] Charlier, R.H., Co-generation from the tides. A review. *Ren. Sustain. Energy Rev.* 2004, **7**, 3, 183–213.

[19] Ministère de l'Industrie et de la Recherche [France], L'usine marémotrice des Chausey. In *La production d'Électricité d'origine hydraulique. Rapport de la Commission de la production hydraulique et marémotrice:* Paris, La Documentation Française [Dossiers de l'Énergie 1976, **9**, IV, 41–49]. [in French: Ministry of Industry and Research of France. The tidal power station of the Chausey {Islands}. In *The production of electricity of hydraulic origin.* Report of the Commission of hydraulic and tide generated energy]; see also R. Bonnefille et al. fn 13.

### 5.3.3 The Anatomy of the Rance River Plant

Besides wiping off the map the local tide mills, construction of the plant required the removal of 1,500,000 m$^3$ of water and the drying up of some 75 hectares of the estuary. As later, though smaller, plants took a different building approach, e.g. avoiding the use of cofferdams, such major civil works are not necessary any longer.

The plant was built in the deepest part of the river. It is a hollow concrete dyke, a tunnel with up- and down-stream linings reinforced by buttresses placed 13.30 m apart. The total width is 53 m with a length of 390 m. The foundations are 10 m below sea level and the top protrudes at 15 m above it. The sas is 13 m wide and 65 m long. The lock gates are vertical shaft sector gates.

Of the 24 bulb groups, sets of twice four groups debit each on a transformer. The transformers are linked by 225 kV oil-filled cables to a substation located on land about 300 m from the plant's western end. Three lines transmit power to Paris, and in Brittany to Aube, Rennes, Landerneau and Brest.

The bulb resembles a small submarine and contains an alternator and a Kaplan turbine .Placed in a horizontal hydraulic duct and entirely surrounded by water, the set functions as a turbine and as a pump. The 5 m diameter alternator is a directly coupled turbine. The 5.35 m diameter turbine is a Kaplan wheel with four mobile blades and vanes; it has a 10,000 W power and rotates at 94 revolutions per minute.

In the direct basin to sea sense, the group furnishes 10 MW for drops of 11 m and 3.2 MW for drops of 3 m. In the sea-to-basin direction, it produces 10 and 2 MW, respectively.

The 14 m wide two-lane highway constructed on top of the dam has put a slow moving and often jammed ferry service out of business and shortened the distance between the towns of Dinard and St Malo by 35 km .The plant has become a tourist attraction. The fishing industry has remained unaffected and a new community has been developed nearby (Gougeonnais). A fish pathway was provided, a noticeable environmental concern for times when such concerns were not front stage considerations. Only one species of fish, the local *lanson*,[alternative spelling *lanzons*] disappeared. No environmental impact study had, however, been made and it took close to fifteeen years before a report was made that concluded on negligible environmental consequences.

Barnier wrote 40 years ago "Apart from being a magnificent technical achievement in itself, the Rance power station is a symbol of innovation in electricity production which offers the great advantage, as compared to the classic hydro-electric power stations, of functioning with perfect regularity"[20].

---

[20] Barnier, L., Power from the tide. *Geographical Magazine,* 1968, **50,** 1118–1125.

## 5.4 The Rance: First and Last of Its Kind?

While Electricité de France[21] has remained convinced of the wisdom and worthtiness of the Rance plant, it has shelved plans for large stations at the Mont-Saint-Michel, the Chausey Islands, the Minquiers. Australia, Argentina, Great Britain are not talking tidal power plant, although Great Britain is funding research and trials. And yet so many advances have been realized. Construction of modules on land, then floated into place has been achieved. Dispensing with the onerous cofferdams is an accepted construction approach. Required large heads (tidal amplitude) have been spectacularly reduced. New turbines have been designed. Small plants seem to answer urgent needs of distant regions. Fears for fisheries and tourism have been allayed. Advantages for ostreiculture have been recorded. The benefits of some modes of operation placed in doubt: are ebb and flow generation, and pumping worth the extra costs? Progress is steadily made with schemes that will allow re-timing by various storage systems.

Tide mills, tidal and tidal current plants' virtues include not to use up irreplaceable stores of energy, not to send into the atmosphere huge—intolerable—amounts of carbon dioxide and other gases, not to contribute to the climate change and the warming up of the atmosphere, and simultaneously to be a continuously available source of energy.

Though some species lost their habitat during the construction period and did not return afterwards, something that will be avoided by dispensing with cofferdams, no "major" biological-ecosystems modifications occurred. Other species colonized the abandoned space. Yet, some environmental changes did take place in the Rance estuary area: sandbanks disappeared, the beach of St Servan, badly damaged during construction, only partially regained its former lustre; high speed currents have developed near sluices and powerhouse in whose vicinity sudden surges occur. The tidal range has been reduced and maxima dropped from 13.5 to 12.8 m while minima increased roughly proportionally. In fact, the green tides (algae proliferation and subsequent strandings) that have been plaguing the French coasts of the Atlantic are a source of far larger economic concern than the implant of the tidal power plant[22].

The economic impact is viewed as favorable with regional industrialization, increase of tourism and substantial growth of the communities involved (St Malo, Dinard).

Forty years went by and except for the far smaller Kislogubsk (Russia) plant, no other large size tidal power plant using bulb turbines has been built. Will there be none ever constructed again? Other ocean energy converting stations seem to have

---

[21] French Government electricity authority. The "State" company was privatized in 2006.

[22] Charlier, R.H. & Lonhienne, T., The management of eutrophicated waters. **In** Schramm, W. & Nienhuis, P.H. (eds) *Marine benthic vegetation. Recent changes and the effects of eutrophication* [Ecological Studies Nr 123]. Berlin-Heidelberg-New York, Springer Verlag 1996, pp 45–78; Charlier, R.H. & Morand, P., Macroalgal population explosion and water purification. A survey. *Int. J. Envir. St.* 2003, **60**, 1 – 13; Morand, P., Briand, X. & Charlier, R.H., Anaerobic digestion of *Ulva* Sp.3. Liquefaction juices extracted by pressing and technicc-economic budget. *J. Appl. Phycol.* 2006.

## 5.4 The Rance: First and Last of Its Kind?

gained favor, waves and winds, and tidal currents. And yet, Korea seems to have opted for a barrage type plant that is to generate 240,000 MW. At this writing, few details have been released even though construction start had been announced for 2006. The plant would incorporate areas of Lake Sihwa [alternative spelling: Siwha] and be built entirely by Korean companies.

Tidal power can indeed also be harnessed by diverting part of a flow and using the to and fro movement of the tide induced current, pretty much like a run-of-the-river approach.[23,24] The flow of the tidal current is diverted in part into a channel where it turns a wheel; some tide mills operated by tapping the energy of the tidal current.[25] Tidal current power is currently more frequently mentioned than tidal barrages.[26,27,28,29] Such power plants need no cofferdams for construction nor barrage, sluices and retaining basin. Though electricity production is more modest, it is far less expensive. Everyone does not agree, however, that tidal currents constitute a big saving from barrage schemes.[30]

Tidal currents vary with depth. In deeper water, density stratification may lead to varied velocity profiles. The stratification may result from salinity, temperature or the amount and nature of sediments transported by the current. Velocity may also be influenced at different depths due to Coriolis effects or even to bottom topography.

---

[23] Charlier, R.H. and Ménanteau, L., The saga of tide mills: *Renew. Sustain. Energy Rev.* 2000.1, 3, 1–44; Charlier, R.H., Ménanteau, L. and Chaineux, M.-C. P. The rise and fall of the tide mill. *Ocean Sciences Bridging the Millennium. Proc. 6th Int. Conf. Hist. Oceanog.* (Qingdao; PRC, Aug. 1998) 2005, Paris, UNESCO- and Qingdao, Ocean Press [1st Inst. of Oceanogr.].

[24] Charlier, R.H., A sleeper awakes: power from tidal currents. *Ren. Sust. Energy Rev.*, 2004, **8**, 7–17.

[25] Wailes, R., Tide mills in England and Wales: *Jun. Inst. Eng.-J. & Record of Transact.* 1941, **51**, 91–114.

[26] Fay, J.A., Design principles of horizontal-axis tidal current turbines. *Proc. Int. Conf. "New Approaches...."*, 1982, No pp. nb.; Kato, N. and Ohashi, Y., A study of energy extraction system from ocean and tidal currents: *Proc. Int. Conf. ECOR '84 & 1st Argent. Ocean Engng Cong. (Buenos Aires)*, 1984, **I**, 115–132; Pratte, D.B., Overview of non-conventional current energy conversion systems: *Prof. Int. Conf. "New Approaches...."*, 1984, Paper N **4**, 1–8; White, P.R.S. et al. A low head hydro head suitable for small tidal and river applications: *Proc.Int. Conf. En. Rural & Island Communities (Inverness, Scotland)* no pp. nb. 1984.

[27] Keefer, R.G., Optimized low head approach to tidal energy: *Proc. Int. Conf. "New Approaches..." op.cit.,* 1982,. [no pp. nb.].

[28] Heronemus, W.E. et al., On the extraction of kinetic energy from oceanic and tidal river currents: *Proc. MacArthur Worksh. Feasib. Extract. Usable E. fr. Florida Curr. (Miami FL)*, 1974, 138–201; McCormick, M.E., Ocean energy sources (class notes): Los Angeles, U. of Calif. at L.A. 1976, p.23; Morrison, R.E., Energy from ocean currents, **in** *Energy from the ocean,* (Library of Congress, Science Policy Research Division-Congress. Res. Serv.) 1982, 95th Congr. Washington, U.S. Gov. Print. Off. pp. 149–173.

[29] Cave, P.R. and Evans, E.M., Tidal streams and energy supplies in the Channel Islands: *Proc. Conf. En. Rural & Island Communities,* 1985, no pp. nb.; *idem,* Tidal energy systems for isolated communities **in** West, E. (ed.), *Alternative energy systems*: New York, Pergamon 1984, pp. 9–14.

[30] Cavanagh, J.E., Clarke, J.H. and Price, R., "Ocean energy systems" 1993, **in** Johansson, T.B., Kelly, H., Reddy, A.K.N. and Williams, P.H., *Renewable energy. Sources for fuels and electricity:* Washington, Island Press & London, Earthscan Chap. 12 p. 523.

Tidal flow being unsteady, the direction and magnitude of tidal currents at depth may, due to bottom friction, lag behind the surface layers' current.

The tidal current resource is gravitational bound, thus highly predictable, but it is, however, not in phase with the solar day. One hundred six locations in Europe, if put "in service", would provide 48 TWh/yr exploitable.[31]

The Rance TPP may not be the last of its kind if the Korean plans come indeed to fruition.

### 5.4.1 The Past and the Future

Some tide mills took advantage of the ebb and/or flood current. Some such mills were even "dual-powered". Tidal current mills operated on one of two systems: they were equipped with a single wheel that rotated with the current between two pontoons, or the mill consisted in a single pontoon with a wheel affixed to each side, similar to the approach with paddle-wheelers.

Tidal current river energy can be tapped both in the sea environment and in tidal rivers and streams. Its potential is large.

A technology assessment conducted by New York University on behalf of the State of New York[32] and dealing principally with the tapping of the tidal current in the East River in New York City, yielded information on a number of devices that could be usable and on the advantages of axial-flow propeller machines.[33,34]

A prototype was installed in the East River's semi-diurnal Eastern Channel in 1985. A new pilot plant has been set up recently.

### 5.4.2 Changes at the Rance TPP

As with other French projects, the Rance River TPP has followed the trend for automation (Ferdinand).

---

[31] European Union-Joule Project, *Exploitation of tidal marine currents* (JOU2-CT94-0355/EC-DG XII): Brussels, European Commission. 1996.

[32] Miller, G., Corren, D. and Franceschi, J., *Kinetic hydro energy conversion systems study for the New York State resource-Phase I, Final Report*: New York, Power Authority of NY State 1982[contract NYO-82-33/NYU/DAS 82–108]; Miller, G., Corren, D., Franceschi, J. and Armstrong, P., *idem-Phase II, Final Report:* New York, Port. Author. 1983 [contract *idem*]; Miller, G., Corren, D. and Armstrong, P., *idem-Phase II and Phase III, Model Testing:* New York, State Energy Research & Development Authority & Consolidated Edison Company of New York, Inc. 1984 [NYU/DAS 84-127]; Miller, G. et al., *idem: Waterpower '85-Int. Conf. Hydropower (Las Vegas NV)* 1984–1985, 12 pp.;

[33] Anonymous, Underwater windmill: *The Energy Daily* December 20, 1984, p.2; Anonymous, East River tides to run electrical generator: *New York Times* Apr. 14, 1985 p.52;

[34] Anonymous, New York plans for hydroplant using kinetic conversion approach: *Bur. Nat. Aff. Energy Rep.* Jan. 3, 1985, p. 9.

## 5.4 The Rance: First and Last of Its Kind?

Changes occurred at La Rance TPP about 1970 and again in 1990. Operators are now assisted by a sophisticated computer that enables to carry out simultaneously automatic control and human shift reduction. The basic principles of the control system have been described by de la Vallée Poussin (1991), Hillariet (1989) and Mourier and collaborators (1990). Automatization insures higher reliability and a more sensitive response to external conditions. The Rance River TPP had its alternators rebuilt around its 30th anniversary (Salvi 1997) and the bulb turbines renovated as well (Nandy 1997).

The AGRA program was put in use at la Rance TPP in 1970. Its new operational model was introduced in 1982. (Merlin 1982, Sandrin 1980). Its special feature include close optimization by dynamic programming, modular program structure, and developed graphical outputs.

But not only have modifications of TPP and of their construction manner, the bulb units used in France, Canada and Russia, have themselves undergone design changes. These are mainly open-flume or in-conduit with hatchcover generator, speed increaser or altenatively variable-speed units. Machinery has also undergone changes: fixed or adjustable blades. fixed or adjustable distributors (Megnint and Allegre 1986, Rabaut 1981). Optimization of the operation was examined in a paper by Orava and Eirola (1983) using the Pontryagin's maximum principle and concomitantly the two-point boundary-value problem, whereof the computational solution was found by the so-called "back and forth method". Bulb units with larger running diameters have been considered. They would improve the economics of TPP (Casacci 1978a,b).

### 5.4.3 Discussion

The performance of the Rance River plant has been closely monitored and this provided the opportunity for the publication of assessments at about decadal intervals. Papers by André, Barreau, Hillariet and Lebarbier deserve mention in this regard[35]. The forty years of operation have proven the viability of tidal power plants.

The Electricité de France, the French Government owned utility responsible for the plant expressed, after 25 years of operation that its specially designed bulb-type turbines performed well. It emphasized, at the time, that the station is not operated to extract maximum energy but, as any commercial plant, to yield the maximum profit.

Yet, by then there had been problems to solve and failures to mend. Some thirty years ago some pinnings were shaken lose, but were repaired successfully. The frequent start-ups wore on the generators, which necessitated repairs. They were provided in the late 1980s with a reinforced coreframe fastener system; this allowed henceforth synchronous starting. Starting is no longer from standstill, and from then on a sequence was followed wherein pumping is preceded by operation as a sea to

---

[35] See fn 13.

**Table 5.1** Relative duration, expressed in %, of the various operation modes at the La Rance Tidal Power Plant

| MODE | % |
| --- | --- |
| Direct generation | 56.00 |
| Reverse generation | 11.70 |
| Direct pumping | 14.70 |
| Reverse pumping | 1.40 |
| Spilling (direct and reverse) | 16.20 |
| *Total* | 100 |

basin sluice. A unit starts at a 30–40% of its rated speed before synchronization with the grid, and stress on the generator is reduced.

Of the six operation modes, generation by basin overemptying has been but rarely used since the early eighties. At that time the company still firmly believed, however, of the value of direct pumping (Table 5.1).

True, plant construction remains capital incentive, but longevity is not an advantage to be passed over lightly. A conventional plant would have shown fatigue. Were a large tidal power plant to be constructed elsewhere several new approaches would be implemented.

**Fig. 5.1** U.S. President F.D. Roosevelt visits Passmaquoddy Tidal Power Plant Project display

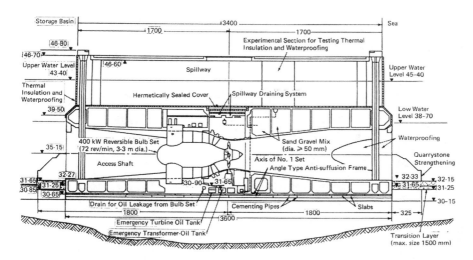

**Fig. 5.2** The bulb turbine which was installed in tidal power plants in France and Russia

**Fig. 5.3** Inside the Range River plant, the bulb turbine system is reversible

**Fig. 5.4** Bulb turbines for high capacity low head power stations

**Fig. 5.5** *STRAFLO*® turbines for low head and tidal power stations

**Fig. 5.6** Rance River plant power station: interior view

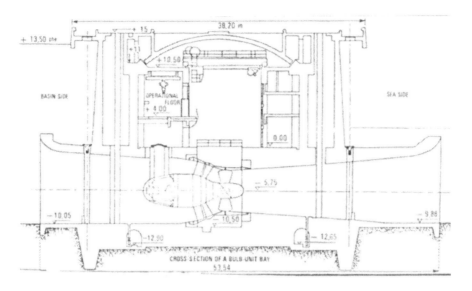

**Fig. 5.7** Cross-section of the Rance River plant power station (St-Malo Fr.)

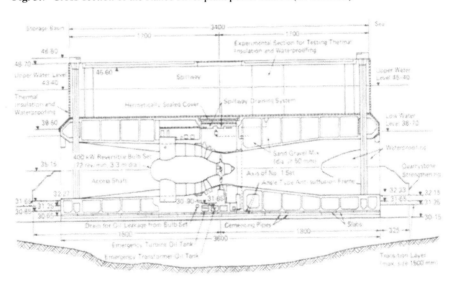

**Fig. 5.8** Cross-section of the Kislaya scheme (Kislaya Bay Rus)

# Chapter 6
# Harnessing the Tides in America

## 6.1 The Quoddy and Fundy Affairs

In North America tide mills had taken a foothold with the arrival of European immigrants. They became gradually obsolete and a few rare remnants subsist. Efforts are underway in New England, but also further south—for instance in New Jersey—to revive, or restore, or even put back to work, many of these mills. In Canada, when it was decided to build a pilot tidal power plant, a replica of the Lequille Mill was made and placed at a short distance of the Annapolis-Royal TPP.

Seven million dollars had been earmarked for the Quoddy undertaking. That was a considerable amount of money for the times (1936) and, in purchasing power, perhaps as much as 20 times that sum by today's dollar value. A start was made with construction of the TPP, but even though some 5,000 persons were once working on the site, the ambitious plans of F.D. Roosevelt never got very far. The "Quoddy" project was to be a joint USA-Canadian undertaking. It ended up as a Canadian plant, nearly half a century later, that, if it eventually got off the ground, was bedeviled by cost overruns, due to excessive charges for the land (Hog's Island) and labor exigencies, better not mentioned here. Nothing was built near Passamaquoddy and a very small amount of the powerful tides of the Bay of Fundy has been generating electricity in Nova Scotia for a bit less than 40 years.

## 6.2 The United States and Tidal Power

Setting aside tide mills, the major step taken in the United States to put the tidal energy to work, is the Passamaquoddy Project. It is, or was, a Canadian-US affair that seems, today, relegated to dusty files as both countries are "going it" their own separate ways. This does not mean that the Quoddy sites are the only locations where tidal energy could be tapped in either country: there are several locations on the Pacific coast.

Three decades before the go-ahead was given to the Electricité de France to launch construction of the Rance River Plant, a visionary American president had dug the first spade into the earth for a similar undertaking in New England. That project has had one thing in common with the rapidly defunct Aber Wrac'h plant: both were started, both were left to slide into dereliction. The French had to stop work because of the Twenties' economic crisis that left the works without money, the US Government laid down the spades because the local electricity companies foresaw an end to their profitable monopoly and pressured the US Congress, through their regional congressmen, to cut off funding for "Passamaquoddy".

Tidal power had been tapped in America long before Franklin Delano Roosevelt came onto the stage. European settlers had brought to the Unites States and Canada their know-how and thus numerous tide mills were built, principally near New York—on Long Island, the Lefferts and Van Wyck mills–, in New England (1617) and in Quebec. News releases of 2007 mention the existence of tide mills in New Jersey as well. Far more than the tide- or sea-mill, the contemporary tidal power plant has to be a sustainable and reliable source of power, and the TPP differs from a run-of-the-river hydro-electric facility in that—until the small head turbines made their apparition and still increased such difference—its hydraulic head is comparatively rather small and very large quantities of water, accurately predictable, will remain available for any foreseeable future.

The Bay of Fundy is a site where tides attain a considerable range, and where, furthermore, the surrounding land configuration offers excellent possibilities for the development of tidal power plants. Here, where the Memramcook and Petitcodiac rivers end their seaward course, the world's largest tides reach ranges of sixteen meters. Tides of only $7^1/_2$ m occur in Cobscook Bay, yet it is in that bay, coupled with Passamaquoddy Bay, thereby providing a $366 \, km^2$ basin, that locations for tidal power schemes have been proposed. It has been a century since various schemes and designs have been studied related to the implant of a TPP in the Passamaquoddy area. Economic rather than engineering factors have put the brakes on construction[1]. A situation similar to that prevailing in Australia existed in the New England States and the Maritime Provinces of Canada: a considerable distance between the location of the potential plant and the actual consumer. Roosevelt saw in the plant an excellent project for his Works Project Administration, providing also jobs, and future development of the region. Since then transmission problems have been resolved, demand for power spectacularly increased, and the utilization of fossil fuels environmentally disastrous and their price reaching un-dreamt of levels.

The General Electric Company, a giant of the electricity industry, developed an interest in 1927, but its plans also came to naught. Yet, the construction costs of just short of $125 million—perhaps as little as one hundredth of the contemporary price tag, would have been retired very long ago[2]. The foreseen production would have reached 1,594 million kW/h.

---

[1] The Canadian pilot plant in the Bay of Fundy is not considered here as a "Quoddy" scheme but rather as a Bay of Fundy plant.

[2] It would have been one tenth of the investment needed half a century later. The one hundredth figure is estimated on the current devalued US dollar.

## 6.2 The United States and Tidal Power

If tide mills operated in the Rance Estuary before the TPP construction, likewise some tide mills had been at work in Passamaquoddy Bay at the end of the 18th century but they had become derelict well before the end of the 19th century. The first applications for a TPP construction in the Quoddy area are associated with the name of Daniel F. Cooper date back eighty years! His proposals dealt with schemes to be located entirely in the State of Maine, others with Canada-US projects[3]. All his applications for Federal Loans were turned down. His final attempt, a two-pool plant, was turned down in 1935 because the US Federal Government had initiated its own project.

Seven million US dollars were allocated, in 1935, by President Roosevelt for the construction, by the US Army Corps of Engineers, of a Cobscook Bay (Maine) TPP. The scheme was to be a single pool plant with a pumped-storage facility near Haycock Harbor and a 22 km long transmission line. Work, started in 1935, came to an end in 1936. It meant also the lay-off of over 4,000 workers. As did the Electricité de France thirty years later near St Malo-Dinard, a special town was created: Quoddy Village to accommodate them. Two dams had already been completed; damaged by a storm, repairs were only halted in 1937. Some forty years after work had stopped, and Quoddy Village been deserted[4], little remained of the Dudley-Treat islands dam but two other dams—carrying a single-track railway line and a State of Maine highway—connecting respectively Corlow and Moose islands and Corlow Island and the mainland (Pleasant Point), subsisted in good condition.

The *coup de grâce* came in 1941 when the US Federal Power Commission spoke out against an exclusively United States plant, finding it economically "uncompetitive" and favoring further exploratory work for a larger international Passamaquoddy scheme. On the other side of the border enthusiasm had also waned—though it picked up again several decades later. The Canadian Government concluded, in a 1945 Report, that a project foreseeing a powerhouse between the Memramcook and Petitcodiac estuaries was uneconomical. Later studies pointed to the Nova Scotia Minas Basin as a more promising site. It is finally there (Hog's Island) that a pilot plant was built (1984 and on).

The 1959 Passamaquoddy Engineering Investigations Report culminated the series of studies undertaken since 1948 jointly by Canada and the United States. It concluded in the Passamaquoddy Fisheries Investigations Report (1959) that though no disadvantage would be suffered by the fisheries industry, the fisheries themselves would probably suffer some adverse effects of the siting of a plant in the Passamaquoddy-Cobscook bays. While herring would be wiped out inside the dam, the economic impact would be minimal since that catch made up only $2^1/_2\%$ of the yield for the St Mary Bay (Nova Scotia)-Cape Elisabeth (Maine) area.

---

[3] 1924, 1926, 1928 and 1933.

[4] The village has totally "disappeared" and the only trace left, in 1970, is a Federal Government owned tract of land.

## 6.2.1 The Passamaquoddy Site

If we go by the standards laid out by de Rouville[5], the site is ideally suited for a power plant. The region has been the locale of crustal and tectonic movements and has undergone alterations caused by intrusions and volcanic flows. Passages around the islands concerned in the project are partially submerged old stream valleys; the islands themselves are parts of pre-Quaternary-glaciation landforms that remained above sea-level. The overburden, in the area, are unconsolidated glacial surface deposits, weathered materials and peat. The bedrock is made up of sedimentary and igneous Paleozoic rocks. We are in a stable region where earthquake risk is very remote. Sand, gravel and crushed rock are exploited and used in road building; they could thus, as is the case with the Rance site, be used as fill in connection with a TPP construction.

An equally favorable hydrological situation prevails. The Quoddy tides are strongly influenced by the resonance of the Gulf of Maine-Bay of Fundy System, although the tidal system in the Bay of Fundy itself is not very resonant. Tidal effects have been the topic of several studies, including some by the Massachusetts Institute of Technology. Wrapping these together it may be concluded that a tidal power complex would modify the in- and out-flow of Passamaquoddy (intended high pool) and Cobscook (intended low pool) bays, and possibly the Bay of Fundy tides, thereby the amount of generated power as well. The MIT reported that the Bay of Fundy tides would, if at all, only increase very slightly. Production would remain satisfactory notwithstanding creation of significant slopes in the pools at Falls Island [lower pool] (Cobscook Bay). The bay's head is favored, and it was frequently mentioned that this site is one of the few in the world where a two-pool scheme can be profitably implanted. A single-pool arrangement was judged more expensive, peak generation rates increased and power produced only intermittently. Other aspects considered include pumped storage and auxiliary river hydro developments.

An Indian tribe, in the area, obtained some funding twenty years ago to build a small TPP. However, hardly any news was ever received on their project since the mid 1980s (Laberge).

## 6.2.2 The U.S. and the U.K.

The United States still shows at best look-warm interest for tidal energy since the Passamaquoddy Roosevelt fiasco of the thirties. Not so the United Kingdom. In May 2005 the Ministry of Trade and Industry awarded £2.7 million (€3.9 million) for the development of "TidEl", a tidal current device.

At the other end of the American continent considerable interest had once been shown to tap the tides of the Gulf of San José.

---

[5] De Rouville, A., 1957, General report on the utilisation of the tidal mechanical energy: *La Houille Blanche-Rev. Intern. de l'Eau* II, 435–455.

## 6.3 Argentina—The San José Tidal Power Plant

The idea of harnessing the energy of the tides to replace the use of coal, that had become suddenly expensive, came to the fore in Argentina in 1923. Further "encouragement" came from the French project to build a tidal power station in Brittany. Work had indeed started at the Aber Wrac'h close to a hundred years ago (1927). The Argentine report, released in 1928, foresaw not one but five plants, Deseado, Gallegos, Santa Cruz and San Julián, with the largest one near San José whose capacity would reach 1,000 MW.

A total of 9,750 GWh—less than the full capacity of the various suitable sites of the Valdes Peninsula—would be produced annually. San José's share would account for nearly half of that (3,650 GWh). The Florentino Ameghino Isthmus separates the gulfs of San José and Nuevo. At approximately 1,000 km from Buenos Aires, the narrow isthmus could, today, play a role similar to that of such peninsulas as the Cotentin (France) discussed in another chapter.

Thus, fifty years ago, a scheme akin to that of the Australians, had been thought of: a canal was to be dug linking the two gulfs, a power plant on the *terra firma*, and barrages constructed by the use of caissons.[6,7] Other proposals came off the drawing boards two years later also suggesting building a canal across the isthmus.[8] That plan proposed an installed capacity of 600 MW and an annual production of 1,750 GWh. De-phasing between moon and solar time could be compensated, it was felt, by combining the TPP project and the river project of El Chocon. Both the 1959 and the 1972 schemes were planned as double-effect (reversible) plants, though the later one also considered a single effect plant. Straflo rather than bulb turbines were to be installed.

Still another possibility was put forth by a State-owned company[9]. This included a dam across the tip of the San José Gulf, a canal crossing the isthmus whereon a 5,300 MW plant, a pumped-storage power plant of 2,500 MW capacity and an upper reservoir of 35 hm$^3$ (262,500 cu.ft) at 100 m above sea level.

The maximum theoretical tidal energy of San José is about 47,200 GWh/yr[10]. A single-basin, double-effect plant would produce 16,000 GWh/yr. Required would be a closure of 13.400 km. The dam's crest would be at 12.5 m above sea level.

Tidal power in Argentina was discussed, at the First Argentine Congress of Ocean Energy, (Buenos Aires, 1984) by E.G. Aisicks and I. Zyngierman ; their conclusions

---

[6] Loschacoff, M.J., 1957, *Un aprovechamiento mareomotriz en la peninsula de Valdes:* Buenos Aires, La Ingeníera.

[7] *Cf.* Kislaya TPP; Delta Plan (Netherlands).

[8] Lemoine, Duport, Preismann, LeMenestrel, Matringe, & Playoust, 1959, *Usina mareomotriz del Golfo San José. Anteproyecto*: Grenoble, SOGREAH.

[9] Anonymous, 1975, *El aaprovechamiento mareomotriz de la Peninsula Valdes*: Chubut, Agua y Energía Eléctrica—Jefatura de Estudios y Proyectos Zona Patagónica.

[10] Bernshtein. L. B., 1965, *Tidal energy for electric power plants—ГИДРОЕНЕРГОПРОЕКТ*: Moscow, State Publishing House. [in Russian but translated by Israel Program for Scientific Translation]; Charlier, R.H., 1982, *Tidal energy*: New York, Van Nostrand Reinhold.

were that the San José TPP would be an economically possible undertaking once cheaper resources have been tapped, possibly exhausted. the TPP would be cost competitive with a relatively low rate of discount and thus studies should not be shelved. The year 2010 was then seen as the time of integration in the national grid. However, it appears that no project is active at present (2008).

# Chapter 7
# Improvements, Adjustments, Developments

No major new technologies are needed for the current construction of tidal power plants, however, it may pay off to foster development and research the interface of a central's output with national grids, calculate a sound estimate of its economic interest, design and site proper implanting, and of course environmental effect and sustainability (Frau 1993). But the 40 years of operation of the Rance and Kislaya plants and the 30 plus years of following the Annapolis facility has taught many lessons. There is nothing sensationally "new", but improvements are numerous, adjustments have been made and development has continued.

## 7.1 Taiwan

Deviating from the traditional tidal power plant, a slightly different scheme was proposed for Taiwan (Jwo-Hwu 1998). It was suggested to generate power from buyoancy and weight at ebb tides. The system apparently can be used anywhere where there is a head, whether a coast, a dam or a pumped-storage reservoir. Perhaps a most significant feature is the ability to shift off-peak generation capacity to the peak need period. The shifting problem (a.k.a. re-timing) has been one of the contentious topics of tidal power generation.

## 7.2 Gorlov's Barrier

As no barrage is needed, the plant requires less capital investment. A "removable barrage" scheme, which involves also re-timing has also been designed by Gorlov (1979, 1980). He proposes to use compressed air to insure re-timing; electricity can be generated directly or compressed air can be stored for later power production. The usual dam is to be replaced by a thin plastic barrier hermetically anchored to bottom and bay sides. A cable stretched across the bay or inlet, attached to several floats, would support the plastic barrier and hold it above water level. Environmental impact would be benign, though a pool of stagnant water may be created behind the barrier; pulled to one side, the barrier would allow evacuation of the impounded water and even navigation.

## 7.3 Japan

The wind power Darrieus turbines have been recently re-examined in Japan for use in tidal current power stations (Shiono et al. 2000). The characteristics of these turbines depend on the strength of rotor and number of blades. Studies recommend, however, that blocking of strait or bay be avoided (Kiho et al. 1998).

As in many areas there is here a major leading interest in tide currents. The Shimada Laboratory of Tokyo's Institute of Technology conducted, in 2005, a successful field test of a 20 kW tidal power generator system. The device included a turbine with a 2.6 diameter, 2 and 3 blades; the rated output is 20 kVA and the rated revolutions 100 rpm. The test showed that the power output is both stable and easily predicted. The output, per area swept, is large and seven times larger than for wind power. A top 17 kW power output was achieved while a daily 57 kWh production attained. Tempering the enthusiastic results of this try-out lasted only four days. It took place in the Ariake Sea, in South Japan, and off the shore near Goto City, sites of a strong tidal currents. More ambitious, yet talked about, is a proposal for a Kuroshio Current Power generation scheme. Ocean current generation units placed every ten km along the Kuroshio, and in the current cross section every km in width would provide as much electrical current as twenty nuclear power plants. Each unit is a 2.5 MW generator (500 kW x 5).[1] Here, however, we are dealing with a marine current, rather than a tidal one. The current has a "width" varying between 70 and 100 km and a depth that averages between 100 and 200 m. Its average (arithmetic mean) speed reaches 1–1 $1/2$ m/s, with a maximum of 2–2 $1/2$ m/s. The proposal suggests placing units every 10 km. No comments are made about the potential effects upon the very active fisheries in the Kuro-Shio, among which sardines and ansjovies.

Still in Japan, the development was reported of a new type of two-way diffuser suitable for a fluid flow energy conversion. Power from such fluid flow is proportional to the cube of the flow's velocity, thence by increasing velocity results in more power. Inspired from experience with wind turbines, a system was developed for tidal streams (Kaneko 2004).

## 7.4 Russia

For countries which have large areas with great tidal amplitudes more voices are raised that favor alternative energy sources over the use of fossil products and nuclear generation (Yakunin et al. 1998). The total world resource for river

---

[1] sakai@iri.titech.ac.jp; http://www.nr.titech.ac.jp~rshimada/

hydro-plants, tidal and wave was estimated, in 1995, at 6,000 GW by Duckers. Of these, 277 Twh of tidal power are found in the former Soviet Union, 20 in Argentina (San Jose Gulf), 17 each in the Severn River location (United Kingdom) and Turnagain Arm (Alaska). The Gulf of Cambay (India) and Cobequid Bay (Canada) could provide respectively 15 and 14 Twh. Bernshtein and Usachev (1997) reminded, on the occasion of the 30th anniversary of the TPP at Kislaya Guba, that Russia has a 17 million kW capacity and could transfer to the combined European power system 50 Twh per year and the proposed Tugur station could add a capacity of 8 million Twh/year usable by the coastal regions of the Russian Federation and Japan. He pointed out that if the American proposal of constructing a combination transport and power tunnel under the Behring Strait would ever materialize, a Penzinskaya TPP could be part of the complex with a capacity of 87 million kW. The potential of China is commonly overlooked, even though it attains 110 million kW. Five hundred bays in seven provinces share this resource: Tchekian, Shantoung, Kouandtong, Iangsou, Hopei and Kianing. Duckers reviewed low head hydroelectric schemes and drew parallels with tidal power plants.

## 7.5 China

Though China has scores of small TPPs in action, and disposes of considerable areas where small or medium-sized plants could be established, it is seldom brought to the fore. A recent "Report" laconically mentions the existence of the plants without indication of most locations nor information on possible future plans.[2]

## 7.6 Great Britain

Taylor (1998) urged for Great Britain the development of nuclear and tidal power as a way to abide by the Kyoto Agreements and to formulate a national sustainable energy policy. And twice the Royal Society Parsons Memorial Lecture stressed the importance of tides generated electricity (Haws 1997, 1999). Compact stations have been recommended (Hecht 1995). The first utilization of tidal power in the United Kingdom predates the Rance: a scheme operated once near Bristol (1931).

Greenpeace spokespersons however had severe words, in 2005, for Britain: "Unfortunately, the debate in the UK has focused more on whether we need a new generation of nuclear power stations. Nuclear power is the epitome of centralized [when we need decentralized networks], outdated electricity generation. Replacing

---

[2] Wu Guihi, 2006, Development status and policy of renewable energy in China: *Renewable Energy 2006*, Proceedings of NEDO [New Energy and Industrial Technology Development Organization] Special Session (Chiba, Japan) paper Nbr 2, 10 pp, no pp.numbers.

existing nuclear stations with new ones would perpetuate the centralized system, entrenching all the costs and inefficiencies that implies"[3].

Since 1999 the United Kingdom has committed approximately £25 million (~$50 million; €33 million) for research and development of wave and tide current devices. Of that amount £5million (~$10 million; ~€6.6 million) for wave and tide testing facilities. The European Commission has approved state aid via the Marine Renewables Deployment Fund.

## 7.7 USA

After 7 years of prototype testing and preliminary studies, state and federal regulators have approved Arlington, Virginia-based Verdant Power's plan to install six underwater turbines in New York City. This array—which could eventually include up to 300 turbines—is presented as being the first grid-connected, multi-turbine source of tidal energy in the world. The statement may be somewhat exaggerated since the Rance River TPP is a multi-turbine facility that has been grid connected for 40 years. Resembling underwater windmills, the approximately 5 m (15-foot) diameter turbines will tap the tidal flow of the East River, in New York, when completed in 2007. The narrow eastern channel of the river moves up to six miles an hour 9.7 km/h, making it one of the fastest water bodies. The sleek, three-pronged turbines swivel to face the oncoming tide, generating up to 35 kw of electricity each.[4]

## 7.8 Norwegian-Dutch Sea and River Mix to Make Power

The Dutch Centre for Sustainable Water Technology or WETSUS, and Norway's independent research organisation SINTEF, working with power company Statkraft, have invented devices that generate electricity by mixing sea and river water. Though it might seem like an exercise in scientific theory destined only for high-tech laboratories, but the process' creators and the European Union, which is funding the Norwegian research, believe the idea's time might have come. Philippe Schild, scientific officer at the European Commission's energy directorate, stated indeed: "There is huge potential in Europe to use this new way of producing electricity, it is a renewable source which does not cause any environmental damage and we think it can play a big role in helping meet our target to increase renewable energy". Though not TPP related, this is reported because it is an ocean-related power-source.

---

[3] Cited in *Renewable Energy World* 8, 5, 37 (2005).

[4] http://www.emagazine.com/view/?2920

## 7.8.1 Co-Generation

Global warming and high oil prices have renewed interest in sustainable energy, with solar, wind, biomass, hydrogen fuel cells, tidal and wave power getting most attention.

But researchers in Norway and the Netherlands, known for their water technology know-how, say there is room for other alternatives given the world's ever-growing appetite for energy.[5]

## 7.9 Some New Ideas

Literature about tidal power has been relatively frequent even if new techniques, new approaches are not numerous. But then there is little needed in new technologies. The lessons of operating a tpp have been drawn and changes in construction approach, turbine characteristics, cathodic protection, the usefulness of pumping, overfilling a basin, bi-directional generation, and so on have been, now for some time, part of the know-how. The multiple basins approach has been shelved and the giga-plants are not really favored, even if Korea seems, nevertheless to go that route.

From Australia have come two ideas, one perfectly logical, the other falling in the group of hard to realize engineers' dreams. Once seriously deliberating whether a TPP in Northwestern Australia would not been a worthwhile undertaking and apparently abandoning that idea for lack of potential power customers, the current proposals focus on the world's busiest maritime highway and on approaches that have worldwide potential. And then there have been a flock of minor proposals that might improve construction or operation.

### 7.9.1 Tidal Delay®

Alexandr Gorlov introduced the "removable dam". Bryden and his collaborators emphasized the advantages of harnessing the tidal current, now comes around an approach baptized Tidal Delay®. It is coupled with a re-timing approach aimed at solving the de-phasing between solar and lunar day, in other words bringing peak demand and peak production closer together.

Tidal power plants require a barrage or a dam. At least the schemes that rely on the tidal up-and-down movement. This explains probably the renewed interest in tapping instead the in- and out-flow movement of the tides, the so-called tidal current.[6] These major engineering works cause, in some sites, silting, thereby necessitating dredging or even leading—as has occurred in China—to the outright

---

[5] http://www.planetark.com/dailynewsstory.cfm/newsid/34044/story.htm

[6] See tidal current power, Chap. 8.

abandon of the tidal power plant. Dams are costly to build even if it has been proven possible to dispense with the expensive cofferdams. They are also viewed askance by several environmentalists. In some instances, due to special geomorphologic and geographical conditions, it is possible also to dispense with barrages.

Tides occur with a certain delay on the two sides of peninsulas, for instance at the Cotentin Presqu'isle in France. In several geographical locations the width of the peninsula is less than at Cotentin and thus even more favorable to a novel approach to putting the tides to work to generate electricity. This is the case in the United Kingdom (e.g. western coast of Scotland and in Wales), in the United States of America (e.g. Alaska on the west coast, Maine on the east coast), on Canada's west coast, India and northwest Australia, where implanting a TPP had once been under consideration.

Australia-based Woodshed Technologies (Pty Ltd) proposed in 2005 to take advantage of the belated arrival of the tide on a given side of a presqu'isle as compared to the other side; this situation creates a "head", to wit a height difference of water level. An isthmus could play the same role as a peninsula.

The "tide" comes in gradually, in other words the water masses take some time to bring two points at the same water level; this delay is the greater if a natural landform (peninsula, isthmus) constitutes an obstacle in the progression path of the tidal current. The water level difference creates a potential energy difference. Points at unequal difference level can be linked by pipes. Placing generator[s] and turbine[s] in these pipes is among the novel approach.

Furthermore the same pipes can be used for the current flow in the opposite direction, thus conceivably generate electricity both at ebb and flood tide. Pipes can be laid on top of the topographic surface or can be buried in a trench. The second alternative is perhaps aesthetically more attractive, though, evidently somewhat costlier as it involved "digging" and "filling". Obviously a significant tidal range and a narrow natural geomorphologic obstacle are needed. Additionally friction and turbulence of and in the water must be kept at a minimum, thence the larger the pipe's diameter that can be accommodated, the better the scheme. Woodshed® has worked with tidal ranges of 1 to 6 m and pipes 100 m–4 km long with diameters of 1–3 m.

As with any TPP even the best site is left on the wayside if the area lacks sufficient customers (demand). Thus the scheme may, in Northwestern Australia, meet the same fate as the "traditional" tidal power plant once considered for the area.

Power production has been calculated for a pipe with a 2 m diameter. According to Hastings' paper for a given pipe length and tidal head difference, a diameter increase of 50% results in a 195% power production increase. He also computed production costs allowing only a 20-year life span for various pipe lengths; thus assuming an unlined 2 m diameter pipe 2 km long and a head of $2^1/_2$ m, a kWh would cost \$0.34. Improvements could result from substituting a siphon to trench and fill engineering, and lining the pipe with plastic or HDPE[7] or fiberglass. The

---

[7] Synthetic piping material.

shorter the pipe the cheaper the power: for a 1 km identical pipe the price drops to less than half ($0.13).

Modules consisting of pipe and a turbine-generator have been proposed, several of which could be placed in a parallel design, but it would be possible to have several pipes debouch into a single, but large, turbo-generator.

## 7.9.2 Where Do "Things" Stand?

A development accord was passed between Woodshed Technologies[8], Lloyd Energy Systems[9] and SMEC Developments[10] to develop a combined harnessing and storing scheme. A United Kingdom pilot plant (1–5 MW) had been scheduled for 2005 but feasibility studies (centered on geographical, financial and environmental aspects were decided upon). It is far more likely that an Australian-British facility will not come on line before sometime in 2009.

The second part of the project deals with power storage. The Lloyd Energy Storage System, which has considerable experience in that technology, is to store and control the Tidal Delay® plants' output. Where SMEC comes in is in the design and construction of the plant and storage scheme.

The UK Department of Trade and Energy granted, in May 2005, €3.9 million (£2.7 million) to SMD Hydrovision for the further development of a tidal current device. The propeller-driven machine, submerged in the sea, is moored to the floor. TidEl is to be a 1 MW full-scale model and grid-connected. Testing is at the tidal test site, near Orkney, of the European Marine Energy Center.

## 7.10 Tapping Channel Tides

Huge, not to say mind-boggling projects, are not an exclusivity of ocean currents harnessing. Russians have dreamt of turning the fluvial giants around and the very serious Electricité de France once toyed with an idea of setting up a mammoth tidal power scheme involving the Chausey Islands. It shelved the plans after many years of discussions; not environmental but financial reasons were given. The tides and tidal current in the English Channel are impressive. Australian researcher G.E. Summer[11] proposed to construct two granite blocks barrage across the Channel and Dover Straits. They would "link" the English ports of Dover and Brighton respectively to Calais and Dieppe in France.

---

[8] See shastings@woodshedtechnologies.com.au ; www.woodshedtechnologies.com.au

[9] See www.lloydenergy.com

[10] See www.smec.com.au

[11] Refer to Dr G.E. Summer 17/42 Macedonia Street, North Haven, 5018 State of South Australia, Australia.

He calculated that no less than 50,000 MW could be produced, even eventually 200–300,000 MW part of which could be "loaded" onto batteries. His plan does not stop there because, if the expected success came to materialize, other such "basins" could be installed incorporating the Channel Islands (Jersey, Guernesey); and to recall in this regard the pre-WWII Mel Project. Conceivably with improved power transfer systems electric current could be fed in grids as distant as the Ruhr industrial basin, thereby reducing carbon dioxide generating, contributing to the climate stabilization and reducing pollution. Seeing no problem in getting appropriate turbines from various manufacturers, he proposed to build the barrage from the seabed up using Bombay-doors equipped granite tankers—already in existence—to dump into place huge blocks.

Summer recognized that such project would be very capital intensive and involve major engineering works. He suggests that tolls that would be levied on ships using the freeways provided through the barrages, would be sufficient to pay the interest charged by lenders that financed construction. Further income would be derived from toll to use the highways built on top of the barrages. One can, however, foresee pressures against such roadways on the part of the operators of Eurochannel and the Channel crossing ferries. And some rumors persist that traffic on the four-lane highway on top of the Rance River TPP are shaking the structure below, adding to the reservations of such use.

Finally Summer, aware of the ships passage problems, would resolve these ensuing problems, passage of ships en route to Rotterdam and Tilbury—he does not mention Dunkirk, Ostend, Nieuwpo[o]rt, Antwerp and some other ports—by creating new docks outside of the barrage as well as locating there oil terminals and pipelines heads. As for ships heading for ports further north, they could pass the barrage at four periods each day when "tidal gates" would be opened at the change of direction of the tide.

This is a *rêve d'ingénieur* as Georges Claude could have dreamt up.[12] The Straits of Dover (Pas-de-Calais to the French) and the [English] Channel have been appropriately called "the busiest sea highway of the world". Ship-owners would raise far more than a fuss about paying a toll where none exists at present, without even thinking of the jam to pass through freeways. An expensive loss of time for shippers, at a time when more than ever time is money. Environmentalists, fishermen and those involved in aqua-farming, possibly conchyliculture people, would also raise major objections; these would range from impact on fisheries, disturbance of benthic and pelagic ecosystems, to climatic and oceanic major disturbances. Summer raised himself the matter of the Gulf Stream wondering if "a warm current could be let in on the southern side of the barrages and still come out on the other side without much change". It could probably be shown that the climatic effects of the Gulf Stream could be changed and this would affect not only biological conditions of the waters, the direction of currents, but as well life on and the economy of the Channel Islands.

---

[12] Claude, G., 1957, *Ma vie et mes inventions*, Paris, Masson.

## 7.11 Turbines

Requirements and aims of turbines for TPPs have changed and bulb, Straflo, Darrieus and other turbines may, in some cases yield their place to other types of turbines. In an effort to reduce research costs, Japanese engineers have examined the possibility of converting turbines designed for tapping wave power into turbines for tidal power.[13] The research was done from the viewpoint of enlarging the possibilities of using tidal power there where only extra-low head is available and the time-varying energy density characteristics are a problem. Reciprocating turbines (impulse and Wells-turbines designed for wave power, cross-flow type Darrieus turbine) were focused for extra-low head tidal power. Similarly gas (petroleum) know-how can be put to use in TPP technology.[14]

A straight wing vertical-axis turbine has been studied by Torii.[15] Less frequently used than other turbines, the hydro turbine generation system may hold promise. An array of turbines in a bay's entrance is effective when uniformly distributed across such entrance. It will nevertheless limit power output, reduce the undisturbed energy flux that normally follows the path. Power generation is compatible if good flushing is maintained even with bay multi-purpose use. Garrett estimates that there is hardly a difference of productivity between conventional schemes where water is trapped at flood tide and then released and continuous operation of current turbines in the entrance.[16]

Kyozuka[17] stressed in an experiment the advantages of tapping tidal flow near a bridge pier. This would also be true near any obstacle to the straight water flow. It is evident that current velocity increases near obstacles. Power being proportional to the cube of the speed of the current, there is more power produced, but the pier also offers the advantage of easy access, thus easy maintenance of both power unit and hydraulic turbine. Model experiments were conducted, using a Darrieus turbine, in a water circulating-tank.

An unusual line of research has been pursued in Japan, a country actively searching for ways to extract energy from the ocean.: why not taking a look at the possibility to adapt reciprocating flow turbines developed for tapping wave energy to tidal energy harnessing? (Takenouchi et al. 2006) This would evidently widen

---

[13] Takenouchi, K., Okuma, K., Furukawa, A. and Setoguchi, T., 2006, On applicability of reciprocating flow turbines developed for wave power to tidal power conversion: *Renewable energy* 31, 2, 209–233; Setoguchi, T., Shiomi, N. and Kaneko, K., 2004, Development of a two-way diffuser for fluid energy conversion system: *Renewable energy* 29, 10, 1757–1771.

[14] Anonymous, 2003, Gas know-how helps harness tidal power: *Int. Gas Engng Manag.* 43, 4, 43.

[15] Torii, T., Ockubo, H., Yamana, M., Seki, K. and Sekita, K., 2006, A study of straight-wing vertical-axis hydro-turbine generation system in ocean environment: *Abstr. & Proceed. Renew. Energy 2006 [Makuhari Messe, Chiba, Japan]* Abstr. p.47, Proceed. pp. 1462–1466.

[16] Garrett, C. and Cummins, P., 2004, Generating power from tidal currents: *J. Waterw., Ports, Coasts. Ocean Engng-ASCE* 130, 3, 114–118.

[17] Kyozuka, Y., Okawa, K. and Wakahama, H., 2006, Tidal current power generation making use of a bridge pier: Abstr. & Proc. Renew. Energy 2006 {Chiba, Japan October 9–13, 2006] Abstr. p. 46, Proc. 1468–1372.

the field of tidal power use with extra-low head and time-varying energy characteristics. Reciprocating turbines with two types of impulse, a wave-energy Wells turbine and a cross-flow Darrieus type one for very-low head hydropower (in a uni-directional steady flow obtained by physical tests models) were compared for their characteristics. This was done in non-dimensional forms; power plant performance with the various turbines, numerically simulated on equivalent scaled turbines; the comparison was based on the low similitude on turbine performance with the non-dimensional characteristics under one of the simplest controls in combination with a suitable reservoir ponds area. "The conclusion was reached that the output of the plant depends [evidently] on the tidal [range] and a pond inundation area."

## 7.12 Re-Timing, Self-Timing

Generating cycles for TPP-s may be self-scheduling so as to take the greatest possible advantage from selling the electricity. The optimisation problem—mix-integer and linear—was solved using commercially available software by de la Torre.[18]

## 7.13 Climate Alteration and Energy Shortage

Global warming and high oil prices have renewed interest in sustainable energy, with solar, wind, biomass, hydrogen fuel cells, tidal and wave power getting most attention.

But researchers in Norway and the Netherlands, known for their water technology know-how, say there is room for other alternatives given the world's ever-growing appetite for energy.[19]

If climate change and the high cost of oil are powerful incentives to search for or develop other methods to generate power, the possible exhaustion of that fossil fuel is an equally strong motive, albeit on the longer term. Assuming the current consumption of oil continues at the same rate, pessimists place the doomsday date in 2040, optimists in 2080, but all seem to agree that the supplies will be down to naught before the end of the current century. There has been talk, decades ago, to exploit the non-conventional oil reserves, *viz.* getting the oil out of tar-sands and oil-shale, but the plans were set aside because the technology then at hand would provide oil at a far from competitive price.

An unusual reaction was triggered by the release of the UN Environmental Panel on Climate Change report that held—for the second time—that human activity strongly impacts the climate, with the combustion of fossil fuels playing a major part (2000). The International Energy Agency predicts, if no measures are taken to

---

[18] De la Torre, S. and Conjo, A.J., 2005, Optimal self-scheduling of a tidal power plant: *J. Electr. Engng*-ASCE 131, 1, 26–51.

[19] http://www.planetark.com/dailynewsstory.cfm/newsid/34044/story.htm

reduce emissions from transportation and power sectors, a still increasing release into the atmosphere of greenhouse gases. It holds that population growth and GNP are proportional to such increase. At the 2002 meeting of the IEEE Power Engineering Society a diametrically opposed view was presented by a group of scientists who, through the paper by Meisen dispute these findings.

## 7.14 Innovations and New Thoughts

Although some authors had expected a resurgence of interest in TPPs (Wilson 1983) after completion of the Annapolis-Royal (Nova Scotia) plant, there was no sudden take-off. Neither did the in depth study of a Severn potential TPP, the feasibility studies for a Garolim TPP (Republic of Korea) and for a Gulf of Kutch (India) plant. It was about the time that a last major discussion of the Chausey Islands Plant (France) found its way in print (Banal 1982, Bonnefille 1982).

Among proposed innovations for TPP schemes, the HVDC[20] bus "comprising transmission lines and converter sub-stations" could be used to "connect tidal power sources and tie interconnected IEPS[21] of the Russian Far East and Siberia" (Belyaev 2002, 2003).

In China the technology of the LonWorks Fieldbus digital communication network has been applied in the 70 KW tidal current electricity generating system. It was thus shown that local devices and working conditions can be managed from an office and that working processes can be logged and adjusted in time. Incidentally the Fieldbus links intelligent local devices and automatic system with multi-directional structures (Li 2003).

Russell, among others,[22] unequivocally states that the benefits of tidal flow have been largely ignored. Indeed, compared to other sources of renewable energy, tidal flow is reliable, predictable, of known gravity and sustainable. It is acceptable from an aesthetic viewpoint, being out of sight, and, contrary to wind power, creates no noise.[23] In San Francisco a sub-sea tidal power station has been under consideration for some time; no moving parts are involved, except on land, advantage would be taken of a venturi, and conventional air-driven turbines would be installed.

It should be more often stressed that tide-generated electricity has no fuel costs, wastes, hazardous by-products, or emissions.

If maintenance is a flaw in the system, there are probably ways to solve this problem, and it does not justify the low profile it has been kept in. Cathodic corrosion protection, another "warning" often uttered, has been improved, witness a study at

---

[20] High voltage direct current.
[21] Interconnected Electric Power Systems.
[22] Anonymous, 2004, Plugging into the ocean: *Machine Design Int.:* 76, 18, 82–90.
[23] Russell, E., 2004, Could the tide be turning? : *Int. Pow. Gener. [UK]* 27, 3, 24–27; Neil, J., Hassard, J. and Jones, A.T., 2003, Tidal power for San Francisco: *Oceans 2004*, 1, 7–9.

the Huaneng Dandong Power Plant (China).[24] This "huge" source of energy should be elevated through research to a much higher profile. Financial requirements remain an impediment but there are also bureaucratic barriers to cope with as well.[25] To be commercially able to fly on its own wings, tidal power must overcome grid connection exigencies, licensing requirements, and not the least marketing costs. POEMS [Practical Ocean Energy Management Systems] made a broad spectrum survey that established agreement across technology and established a list of objectives that will push ocean energies, including tidal of course, towards competitive commercial viability.[26] Switching from building a wind farm in Victoria and from pine and cane use in power stations, Verdant (an Arlington VA company) would like to address commercial viability of tidal power with a plant in New York City's East River.[27]

Schemes have been investigated to convert marine energy into electricity using, instead of pneumatic, hydraulic, et al. linkages, direct drive electrical take-off. Mueller and Baker investigated a linear vernier hybrid permanent magnet machine and an air-cored tubular permanent magnet machine.[28] Two main types of generators have been introduced recently, to wit the seafloor sited and the hydroplane. Some funding is expected from the British government. Gorlov's new helic[oid]al turbine is "environment friendly".[29] The unidirectional rotation machine is particularly suited for reversible use in tidal streams, bays, estuaries, etc, but also in rivers and needs neither dams nor canals.

Already in 1993 additional research had been suggested in the matter of the interface of the TPP output with national grids, that a sound assessment of such plants, their economic interest, design, and implementation according to the site, and environmental effects was both desirable and urgent (Frau).

Tidal power received a major examination during the 1993 meeting of the IEEE. The paper by Hammons discussed funneling, reflection, and resonance phenomena. He found that energy extracted is proportional to the amplitude while the needed sluicing area is proportional to the square value of tidal amplitude.

---

[24] Liu, G-c., Wang, S-d., Song, S-j., Zu, Y-j., Sun, L., 2004, Corrosion and control of sea water circulation system of Huaneng Dandong Power Plant: *Electric Power* 37, 2, 87–88.

[25] Wood, J., 2005, Marine renewables face paperwork barrier: *IEE Review* 51, 4, 26–27.

[26] Anonymous, 2003, Ocean energy development. Obstacles to commercialization: *Ocean 2003* 4, 2278–2283.

[27] Grad, P., 2004, Changing tide of power generation: *Eng. Australia* 76, 11, 51–52.; Gross, R., 2004, Technologies and innovations for system change in the UK: status, prospects and system requirements of some leading renewable energy options: *Energy Policy* 32, 17, 1905–1919; Anonymous, 2003, Ocean energy development. Obstacles to commercialization: *Oceans 2003. Celebrating the past. Teaming toward the future* 4, 2278–2283.

[28] Mueller, M.A. and Baker, N.J., 2005, Direct drive electrical take-off for offshore marine energy converters: *Proc. Inst. Mech. Eng. Assn, J. Power Energy [UK]* 219, 3, 223–234.

[29] Flin, D., 2005, Sink or swim: *Power Eng.* 19, 3, 32–35; Fox, P., 2004, Designs on tidal turbines: *Power Engineer* 18, 4, 32–33; Gorlov, A.M., 2003, The helical turbine and its applications for tidal and wave power : *Oceans 2003*, 4, 1996.

## 7.15 Public Acceptance

To reduce emissions the UN Development Program, but also the Royal Dutch Shell Company long range planners, suggest the use of renewable energies. The price of these, however, must be made competitive with that of conventional fuels. Be it mentioned here that the steady rising cost of oil, automatically contributes at narrowing the price gap. Another condition is that means be found to secure systems' reliability, as intermittent resources, such as tidal power among others, expand into the daily load mix. And there is a psychological factor as well, linked to the final recognition by governments—particularly the US—to wit acceptance by the public at large and the governments' willingness to address the problem of climate change.

The World Energy Council sees excellent opportunities between the present and the end of the next decade. Waiting for the end of the next decade for tapping renewable energy sources is perhaps an ill-advised gamble with the environment. A rapid turnover is excluded because of investments in traditional facilities. The choices that are made today, however, will determine whether the cortège of ill consequences of climate warm-up, will be blocked by the development and implementation of better technologies, or will proceed with its scientific uncertainties, an Absalom sword.

The concern about environmental effects of a barrage led to the Hastings (2005) proposal. The difference in time of tides between the two sides of the Cotentin Peninsula (France) has retained the attention of geographers and engineers alike. It was mentioned earlier, The tidal phase difference, or tidal delay, could be put to use, and the landform constitute a natural "barrage". The Australian Woodshed Technologies proposes to utilize the phenomenon to capture tidal energy. Still another scheme to do away with the TPP barrage has been put forth by Appleyard (2005). The hydro-electricity generator of Hi-Spec Research (Fowey, UK) is an off-shore system that has a series of four channels created from two outer walls and three rows of chambers housing turbines.

## 7.16 New Technologies

A new perspective is perhaps added to utilization of tidal stream (current) by Mueller and Baker (2005) in their analysis of conversion to electricity of energy produced by marine energy converters, using direct drive electrical power take-off (cf. 7.14 p. 138). Taking into account the electrical properties of the topologies of these machines, problems of sealing, lubrication and corrosion were studied.

Even though 2006 dated information pertaining to tidal energy use in Russia indicates a current disinterest, Usachev et al. published in 2004 both a review of its large-scale use and the observations made over 35 years of operation of the Kislaya Bay TPP. The TPP has operated successfully, with moving parts underwater, under extreme climatic conditions. The plant has been used as a biological test site in a nearly closed-off basin, including fisheries and mariculture observations. A variety of units were tested among which cathodic behavior, sorption and electrolytic. The reporters confirmed that a TPP can be integrated into a power system, in basic and peak periods of the load diagram, allowing for the specific conditions of tidal energy generation.

Trials to tap tide current have been conducted in the New York East River (Anonymous 2004).

## 7.17 Wrap-up

An attempt at defining the viability of tidal energy generated power was made as recently as 2005 (Wood), however the rise in oil prices has not been fully taken in account and a revised schedule is timely.

The European Commission funded CAOC[30] which put in place funding and a platform enabling devices developers, wave and tidal energy researchers, and standard agencies to share information and knowledge that will facilitate the transition of wave and tidal energy from an energy research technology to one approaching commercial competitiveness with a medium-term time frame (Johnstone et al. 2006).

The role of hydropower for developing countries is repeatedly emphasized. Small scale operations of tidal currents will furthermore bring electricity to remote locations. This should not obliterate the fact that more hydropower could be used in industrialized lands. Review papers abound and stress—now that marine winds power is well established—the potential, in that order, of wave, tides and tidal streams. Kerr (2005) estimates that such marine sources could contribute, in the UK alone, some 75 TWH underscoring—once again—that integration of such power into the electricity grid remains a critical issue. Efforts would be also welcome in the domain of tidal regimes: how can energy be extracted with the smallest possible disturbance to the regime?

There is currently a flurry of news releases focusing on tidal power (Anonymous 2002a,b, 2003, 2003a,b[Norway], 2005, Bryden 2003). One may expect the trend to perdure, and perhaps the dream will come through elsewhere than in France, Russia, Canada and China. Which by no means signifies that everyone is in tune with the chorus: for Spare (2003) tidal power is a wash-out!

### 7.17.1 Does the CEO Get a Pass?

Charlier asked ten years ago whether tidal power had come of age (Charlier and Justus 1993); one may query whether, in general, does Conversion of the Energy from the Ocean get a pass? It is remarkable that the problem of tapping ocean

---

[30] Co-ordinated Action on Ocean Energy.

## 7.17 Wrap-up

energies remains resilient for so long. The matter bounces back periodically. Yet, positive action is limited.

Had there been such hesitation about railroads and aviation where would we be?

It took more than 150 years to finally decide to construct the Rance River plant; prior to this historic move of Charles de Gaulle, ailing tidemills were dying off, among others a timid tidal power plant had briefly functioned in Boston Harbor, another in Great Britain, still another in what is now Suriname, and, loudly heralded attempts on the US and French coasts aborted either for political reasons or were grounded by the financial crisis.

Three stations are functioning: in France, Nova Scotia and Russia. There is a growing literature about China's numerous mini-plants but very little is actually known about them (Charlier 2001). There is no shortage of announcements coming from China, Korea and Russia about, mostly large, new plants. But these remain *pronunciamentos* at international congresses and in press releases, so far all *sans lendemain*. France shelved its mega-project of the Chaussey Islands, but then the Electricité de France has been on a nuclear binge. Environmentalists raise the same fears of disturbed ecosystems, yet not so-environmentally-benign river barrages are built.

The electricity production costs calculations sometimes favor the tidal plant, or are said to be about the same as those of conventional or nuclear plants, with the conventional stations $CO_2$-polluters. Capital costs remain high for a tidal power plant but the longevity of the tidal plant is given at 75 years compared to 25 for a fossil fuel thermal central and between 30 and perhaps 40 years for a nuclear one.

Pleas to harness tides are more frequently and forcefully voiced than before. In these authors' view, whatever the future of nuclear generation, and probably of thermal plants, tidal power, and other ocean sources ought to be added as at least complementary generators to quench man's thirst for kilowatts.

# Chapter 8
# Current from Tidal Current

## 8.1 Introduction

With at least sixty major papers in print and several international conferences on the subject, over the last decade, it appears that, contrary to assertions made in some trade-and-news journals, interest in tidal power—using barrages—is far from being on the wane.[1,2] This is particularly true in the Far East, *viz.* China, Japan, Korea.[3]

Tidal power can be harnessed, as did tide mills, by creating a retaining basin and using the up and down movement created by the tides, or by diverting part of a flow and using the to and fro movement of the tide induced current, pretty much like a run-of-the-river approach.[4] The flow of the tidal current is diverted in part into a channel where it turns a wheel: some tide mills operated by tapping the energy of the tidal current.[5] Substitute a turbine, place it in the dam and you have a tidal current energy plant.

In discussions dealing with extraction of tidal energy, tidal current power has been far less often mentioned that tidal barrages yet, this "sleeper" has been

---

Passages from this chapter are excerpted from an article that appeared in *Renewable and Sustainable Energy Reviews*. Elsevier Science, the publishers of the periodical have kindly granted permission to reprint them.

[1] Charlier, R.H., 2002, *Resurgimento* or *Aggiorrnamento?* Tidal power at 30 years: *Renew. Sustain. En. Rev.* **I**, 4, 271–289.

[2] Charlier, R.H., 2003, Sustainable co-generation from the tides. A bibliography: *Renew. Sustain. En. Rev.* **7**, 4, 189 213 [Part 3];Charlier, R.H. and Ménanteau, L., 2000, The saga of tide mills: *Renew. Sustain. En. Rev.* **1**, 3, 171–207; Charlier, R.H., Ménanteau, L. and Chaineux, M.-C. P.; 2003, The rise and fall of the tide mill: *Proc. 6th Int. Conf. Hist. Oceanog.* (Qingdao; PRC, August 1998), publ. by UNESCO-Paris and 1st Inst. of Oceanogr. Qingdao-PRC. Part III, chapt. 39.

[3] Charlier, R.H., 2001, The view from China—Small is beautiful. *Renew. Sustain. En. Rev.* **5**, 3, 403–409.

[4] Charlier, R.H. and Ménanteau, L., 2000, The saga of tide mills: *Renew. Sustain. En. Rev.* **1**, 3, 1–44; Charlier, R.H., Ménanteau, L. and Chaineux, M.-C. P.; 2003, The rise and fall of the tide mill: *Proc. 6th Int. Conf. Hist. Oceanog.* (Qingdao; PRC, Aug. 1998), publ. by UNESCO-Paris and 1st Inst. of Oceanogr. Qingdao-PRC. Part III, Chapt. 39.

[5] See fn 4; Wailes, R., 1941, Tide mills in England and Wales: *Jun. Inst. Eng.-J. & Record of Transact.* **51**, 91–114.

considered as a power source with steadily increased frequency for better than a decade.[6,7,8,9] As one may dispense with cofferdams, barrage, sluices and retaining basin, a tidal current power plant can be built and provide electricity far less expensively. Electricity production is of course more modest than with a "barrage plant". The view that tidal currents constitute a big saving compared to barrage schemes is not, however, held universally.[10]

An advantage of the tidal current resource is that being gravitational bound, it is highly predictable; it is, however, not in phase with the solar day. One hundred six locations in Europe if put "in service would provide 48 TWh/yr exploitable.[11]

## 8.2 Tidal Current

The tide phenomenon is the periodic motion of the waters of the sea—and is observed upstream of several rivers—caused by celestial bodies, mainly the moon and the sun. The tide results from the gravitational pull and the earth's rotation. Tide and tidal currents must be differentiated, for the relation between them is not simple, nor is it everywhere the same. In its rise and fall the tide is accompanied by a periodic movement of the water, the tidal current; the two movements are intimately related.

The current experienced at any time is usually a combination of tidal and nontidal currents (cf. p. 67). Offshore, the direction of flow of the tidal current is usually not restricted by any barrier and the tidal current is rotary.

---

[6] Fay, J.A., 1982, Design principles of horizontal-axis tidal current turbines. *Proc. Int. Conf. "New Approaches...."* No pp. nb.; Kato, N. and Ohashi, Y., 1984, A study of energy extraction system from ocean and tidal currents: *Proc. Int. Conf. ECOR '84 & 1st Argent. Ocean Engng Cong. (Buenos Aires 1984)* I, 115–132; Pratte, D.B., 1982, Overview of non-conventional current energy conversion systems: *Prof. Int. Conf. "New Approaches...."* Paper No 4, 1–8; White, P.R.S. et al. A low head hydro head suitable for small tidal and river applications: *Proc.Int. Conf. En. Rural & Island Communities (Inverness, Scotland)* no pp. nb.

[7] Keefer, R.G., 1982, Optimized low head approach to tidal energy: *Prof. Int. Conf. "New Approaches..." op.cit.* [no pp. nb.].

[8] Heronemus, W.E. et al., 1974, On the extraction of kinetic energy from oceanic and tidal river currents: *Proc. MacArthur Worksh. Feasib. Extract. Usable E. fr. Florida Curr. (Miami FL)* 138–201; McCormick, M.E., 1976, Ocean energy sources (class notes): Los Angeles, U. of Calif. at L.A. p.23; Morrison, R.E., 1978, Energy from ocean currents, **in** *Energy from the ocean*, (Library of Congress, Science Policy Research Division-Congress. Res. Serv.) 95th Congr. Washington, U.S. Gov. Print. Off. pp. 149–173.

[9] Cave, P.R. and Evans, E.M., 1985, Tidal streams and energy supplies in the Channel Islands: *Proc. Conf. En. Rural & Island Communities* no pp. nb.; idem, 1984, Tidal energy systems for isolated communities **in** West, E. (ed.), *Alternative energy systems*: New York, Pergamon pp. 9–14.

[10] Cavanagh, J.E., Clarke, J.H. and Price, R., 1993, "Ocean energy systems" **in** Johansson, T.B., Kelly, H., Reddy, A.K.N. and Williams, P.H., *Renewable energy. Sources for fuels and electricity:* Washington, Island Press & London, Earthscan Chap. 12 p. 523.

[11] European Union-Joule Project, 1996, *Exploitation of tidal marine currents* (JOU2-CT94-0355/EC-DG XII): Brussels, European Commission.

## 8.3 Energy Potential

The tidal current is the rotary current that accompanies the turning tide crest in the open ocean and becomes a reversing current, near shore, moving in and out, respectively as flood and ebb currents. There is an instant or a short period—the slack period—when there is little or no current, at each reversal of current direction. During the flow in each direction current speed varies from naught to a maximum—strength of flood or ebb—about midway between the slacks. The shorewards, and upstream, movement is the flood and the seawards, and downstream, movement is the ebb.

Both the rise and fall of the tide, and the flood and ebb of the reversing current can be harnessed to produce mechanical and/or electrical power. Tidal currents are alternating and their maximum velocity occurs at high and low water. The motion is uniform from surface to bottom, except for wave interference at the surface and increases with distance. Because of superimposition by other currents, observation of tidal currents is difficult and requires extensive complex data.

Tidal currents may be semi-diurnal, diurnal or of mixed type, corresponding largely to the type of tide at the site, but often with a stronger semi-diurnal trend. The most common type is a to a greater or lesser degree, mixed one.

Treating tides as waves, a progressive tide wave will have a shallow wave horizontal orbital velocity ($U_s$) given by (8.1) wherein A is the wave amplitude, $\sigma$ the angular velocity of a particle undergoing a circular motion as the tide wave passes by, k the wave number, h the water depth in which the wave is progressing, x the distance from a point of origin and t the time from a particular instant:

$$U_s = A\sigma/kh \cos(kx + \sigma t) \tag{8.1}$$

Tidal currents are an appreciable energy resource in relatively shallow water, near continents. When particular geometry comes into play, bottom and sides may impede the flow and speeds from 9 to 19 km/h have been registered.

## 8.3 Energy Potential

Flow-of-the-river potential is directly proportional to elevation above sea level and precipitation run-off. If a "sector" is the distance between two successive confinements—about 10 km—the linear potential of a river is given in the eq. (8.2)

$$P_f = 9.8 x Q_m x H \tag{8.2}$$

wherein H is the elevation difference, expressed in meters, above sea-level, between points of origin and exit of a sector, $Q_m$ the mean discharge, expressed in m$^3$/s, at the end point of a section, and $P_f$ the mean power expressed in kW. By summing the successive $P_f$ values ($\Sigma P_f$) a river's "potential" can be calculated.

The theoretical energy ($E_f$) is given by (8.3):

$$E_f = 8760 \Sigma P_f \tag{8.3}$$

## 8.3.1 Regional Potential

To calculate $P_{f\,(region)}$ or linear potential for a specific region, requires to know the elevation above sea-level of the individual basins ($H_i$) and the mean run-off of several basins; an estimate of the theoretical linear value of the per annum potential, in millions of kWh (kWh.$10^6$), is now possible ($P_f$). Taking $H_{med}$ as the median of the $H_i$ values, and V being the precipitations' run-off, expressed in millions of m$^3$, the potential is found through (8.5):

$$p_f = \frac{9.8 \times 8760 \times Q \times H}{8760 \times 360} \tag{8.4}$$

$$P_f = V \times H_{med}/367 \tag{8.5}$$

A region's potential can thus be given in kWh/km$^2$, but, particularly in less developed countries, it often is not precised due to the lack of long-term hydrological data. As usual economic potential is lower than the theoretical potential for such reasons as several streams having too small a discharge or waters being diverted for irrigation. In some cases not the "value" but environmental constraints—natural or social—may make the implantation of a generating station unacceptable. The ratio between two values may well vary around 1–5. Finally, the economic potential value may vary in time.

Another estimation of extractable power ($P_x$), calculated in W/m$^2$, is given by (8.6) in which typical values of the extraction efficiency factor ($\mu$), the velocity profile factor ($K_s$) and the spring/neap tide factor ($K_n$) are respectively 0.25, 0.424 and 0.57; w is the fluid density in kg/m$^3$.

$$P_x = 1/2 \mu w K_s K_n V^3 \tag{8.6}$$

$$P_x = 0.3 w V^3 \tag{8.7}$$

## 8.4 Geographical Distribution of Promising Sites

Major tidal currents are encountered in the Arctic Ocean, the English Channel, Irish Sea, Skagerrak-Kattegat, Hebrides, the gulfs of Mexico and of St Lawrence, the Bay of Fundy, such rivers as the Amazon and Rio de la Plata, the straits of Magellan, Gibraltar, Messina, Sicily, Bosporus. The tidal range is observed as far as 800 km upstream on the Amazon River! In the Far East currents are encountered, a.o. near Taiwan and the Kurile Islands. Northwest and Western Australia have their share. There are many other locations.

Most commonly cited examples are the Pentland Firth, Irish Sea North Channel, Alderney Race, Isle of Wight to Cherbourg, Orkneys to Shetlands. The Florida

Current has been mentioned repeatedly as it is assumed it could provide 25 GW, but the idea of tapping it has caused more than a raised eyebrow among environmentalists.

Where narrow straits occur between landmasses or are adjacent to headlands, large tidal flows develop. For instance in the Iroise Sea, off the Brittany coast, current speeds of 8 knots are not infrequent in the Fromveur passage.

Very high one-way tidal currents exist in the Far East Indies and at the southeast of New Guinea-Papua. Indeed, the westerly movement of the "planetary tidal wave" increases substantially the southwesterly ebb flows in the connecting Pacific-Indian oceans channels, with ebb tides reaching in spots 10 knots. An exploitable 70 TWh/yr would be available in the Sibulu Strait of the Philippines.[12]

Even if huge amounts of energy are available it seems that tidal current power is best adapted for regional, even local sites.

## 8.5 Proposed Schemes

Twenty years ago, it was felt that a scheme most suitable to attain an acceptable, favorable benefit to cost ratio ("rentability") would be one in which rotors would be anchored, but suspended in mid-water—precisely to avoid wave influence—and let drive hydraulic pumps, while conversion to electricity would occur at a central facility servicing several rotors; if these were spaced over some distance, the de-phasing due to tide variation would be compensated *in partim*. More turbines could be inserted in the system, an idea based on the belief that cost would be rather low.

Fifteen years ago, it was suggested to anchor in a line a series of floating turbine and generator units along the flow of the tidal current. A Savonius-type rotor might fulfill the role, however, due to the large size needed, a string of units would have to be stretched over at least half a kilometer; if the rotors drove, instead, hydraulic pumps a hydraulic motor could combine each rotor's output, but then we would be back at the 1980s proposal. Finally, as an alternate turbine-generator, a horizontal-axis free-stream machine could be used, though with a nine-meter diameter required.

Musgrove has felt that the most straightforward tapping scheme would be an underwater equivalent of a windmill. So has Heronemus.[13] Musgrove's scheme used vertical rotors.

## 8.6 Glance at the Past and Look into the Future

The tidal current was used by water mills on Evrepos Strait, in Cephalonia, in the floating tide mills on the Danube, Tiber, Seine and Russian rivers. A plant func-

---

[12] Carstens, T.A., 1998, *A global survey of tidal stream energy*: Oslo-Norway, SINTEF.
[13] See fn. 15.

tioned briefly in northwestern Iceland and another has been mentioned for the Faroë Islands. The Danube tide mills used undershot wheels since Roman times to harness the tidal current. Some of them were still in use below the Iron Gates as late as 1970.

### 8.6.1 The Modest Forerunners

The term water mills has commonly designated run-of-the-river mills situated on waterways where there was/is no tidal current. The terms sea mill, and later tide mill, designated mills that took advantage of the tides with or without retaining ponds. There were thus tide mills that took advantage of the ebb and/or flood current. Some such mills were even "dual-powered". Tidal current mills operated on one of two systems: they were equipped with a single wheel that rotated with the current between two pontoons, or the mill consisted in a single pontoon with a wheel affixed to each side, similar to the approach with paddle-wheelers.

The Dunkirk (Dunkerque, France) "Perse mill" (end of 17th century to 1714), the Bacalan mill a few kilometers north of Bordeaux on the Gironde River, the El Ferol mill (Galicia, Spain), used an ingenious hydraulic machinery that allowed them to use both ebb and flood currents for power production.[14] So could the scheme installed in the Thames River under London Bridge. The Demi-Ville (Morbihan Department, France) was an example of dual-powered mill using both the fluvial current and the tidal currents.

### 8.6.2 The Contemporary Scene

Tidal river energy can be tapped both in the sea environment and in tidal rivers and streams. Its potential is large and a mere 10% of the energy in Great Britain was estimated sufficient to provide more than 5% of that country's electrical needs a quarter of a century ago.[15] The 8-knot current of the river underneath the Golden Gate Bridge (San Francisco) can provide all the bridge's needs in electricity. Likewise were the Florida Current to be harnessed 25 GW of electricity could be produced. An "aqua power barge", capable to "harvest" energy along coasts and on tidal rivers, proposed in 1979, would use a high-impulse low-head turbine; with a 6 knots current 50 kW of installed power could be produced.

Patents have been taken out in the United States since the 19th century for a variety of devices intended to tap directly the energy of waterways; they encompass small units as well as "giant" paddlewheels. AeroVironment Inc., where the Coriolis Project was developed[16], examined the river energy resource for the Western

---

[14] Tascón, I.G., 1987, *Fábricas hidraulicas españolas:* Madrid, Minist. Obr. Publ. P. 225.

[15] Musgrove, P., 1979, Tidal and river current energy systems: *Proc. Conf. "Future Energy Concepts", Inst. Elec. Eng. [London],* IEEE Conf. Publication No 17, 114–117.

[16] A project that examined a scheme to tap the Florida Current.

## 8.6 Glance at the Past and Look into the Future

United States, the economics of ducted and un-ducted axial flow turbines and even carried out some small-scale rotor model tests.[17]

Davis and Swan sought to develop a ducted Darrieus design.[18] Designs of non-conventional conversion systems have been frequently reviewed (Pratte, Davis, and others).[19] Vertical axis turbines were proposed by Davis and Swan.[20]

A technology assessment conducted by New York University on behalf of the State of New York[21] and dealing principally with the tapping of the tidal current in the East River in New York City, yielded information on a number of devices which could be usable and on the advantages of axial-flow propeller machines.[22] The various types of KHECS[23] included waterwheels, free-ducted and Wells rotor axial flow turbines, Darrieus, Savonius, and cyclo-giro type vertical axis rotors and the Schneider Lift Translator. The conclusion of the studies was that the system would cost less than $1,700/kW installed.[24]

A prototype was installed in the East River's semi-diurnal Eastern Channel in 1985.

Attached to the side of a bridge the 4.3 m diameter device used a three-blade conformal design. Modest ducts had been attached to the screen hoop to test their potential cost-effectiveness. The unit was dismantled for inspection after a short period of operation.

Though hardly tidal current schemes, many proposals have been ventured to link various seas, streams and canals. Some visionaries, including Theodore Herzl in 1902,[25] have suggested a canal linking the Dead and Mediterranean seas, and

---

[17] Radkey, R. and Hibbs, B.D., 1981, *Definition of cost effective river turbine design. Final report for the period September 30, 1980–December 31, 1981. AeroVironment; Inc.*: Washington DC, Department of Energy (DES82010972).

[18] Davis, C.V. and Swan, A.H., 1982, Extracting energy from river and tidal currents using open and ducted vertical axis turbines. Model tests and prototype design: *Int. Conf. "New Approaches to Tidal Power" Proc.* (Bedford Inst. Oceranog., Dartmouth, NS) [no pp. nb.]; Davis, B. and Swan, D., 1983, *Vertical axis turbine economics for river and estuaries in modern power systems:* Montreal, Nova Energy Ltd.

[19] see fn 17; Pratte, see fn 6; Davis and Swan see fn 17.

[20] See fn 17.

[21] idem.

[22] Miller, G., Corren, D. and Franceschi, J., 1982, *Kinetic hydro energy conversion systems study for the New York State resource-Phase I, Final Report*: New York, Power Authority of NY State [contract NYO-82-33/NYU/DAS 82–108]; Miller, G., Corren, D., Franceschi, J. and Armstrong, P., 1983, *idem-Phase II, Final Report:* New York, Pt. Auth. [contract *idem*]; Miller, G., Corren, D. and Armstrong, P., 1984, *idem-Phase II and Phase III, Model Testing:* New York, State Energy Research & Development Authority & Consolidated Edison Company of New York, Inc. [NYU/DAS 84–127]; Miller, G. et al., 1984–1985, *idem: Waterpower '85-Int. Conf. Hydropower (Las Vegas NV)* 12 pp.; Anon., 1984, Underwater windmill: *The Energy Daily* Dec. 20, 1984, p.2; Anonym., 1985, East River tides to run electrical generator: *New York Times* Apr. 14, 1985 p.52; Anon., 1985, New York plans for hydroplant using kinetic conversion approach: *Bur. Nat. Aff. Energy Rep.* Jan. 3, 1985, p.9.

[23] Kinetic hydro energy conversion systems.

[24] In terms of 1980-$.

[25] In his novel *"Altneuland"*.

proposed to tap the current to generate electrical power. Some thoughts to that effect had been expressed as early as 1850. It was however the James Hayes Commission which, in 1943, made a first assessment. An Israeli commission recommended moving ahead with a linking project at the end of the 1970s.[26] Apparently the plan has been laid to rest, probably the better so in view of the probable ecological consequences it would have.

The advantages of the *Turbodyne Generator* were praised in 1982: the amount of turbine material is small and the high speed vertical-axis turbine was shown in theory and actual tests to perform highly with low heads; power generation is possible, in ebb and flow tides, independently of current speed, provided the current has a small head (even <1.5 m); silting risks are low, environmental impact rather benign and no sluices are necessary.

Baker and Wishart conducted a study covering three small estuaries and seventeen sites in Great Britain and, in terms of 1983 dollars, arrived at a cost varying between $6.10 and $6.30, depending on the number of turbines, per kWh. The cost (C) of a barrage is given in (8.8), which includes correction factors for shallow margins and ranges ≤ 1

$$C = \frac{L^{0.8}(H+2)^2}{A(R-1)} \quad (8.8)$$

wherein L is the length of a barrage, H is maximum depth, A is the basin area and R is the tidal range.[27]

Among recommended sites were the Camel River (Cornwall), the Taw-Torridge estuary (Devonshire), Milford Haven (South Wales), Loughor Estuary and several on the Mersey River.[28]

The Salford Transverse Oscillator could harness energy from tidal currents a.o. in rivers and tidal inlets; it could function in basins as small as 0.5 km$^2$, e.g. Loch Heuran (Scotland). Installation of a prototype was being considered in 1993. If P is power, $\omega$ the specific weight of water, Q the water discharge, H the head, then

$$P = \omega Q H \quad (8.9)$$

When the flow reaches as little as 0.49 m/s an immersed Savonius type rotor driving a generator could power a marine beacon, and greater efficiency could be

---

[26] Dolan, R., 1984, The Dead Sea-Mediterranean Canal project: *European Scientific Notes* **38**, 5, 262–264; Kaplan, R.D., 1981, New Canal will link Mediterranean and Dead Sea: *The Christian Monitor* April 13, p.5.

[27] Baker, A.C. and Wishart, S.J., 1986, Tidal power for small estuaries: *3rd Int: Symp.Wave, Tidal, OTEC & Small Scale Hydro Energy "Water for Energy", Proc.* Paper No 9, 115–123; Wishart, S.J., 1981, A preliminary survey of tidal energy from 5 UK estuaries: *Proc. 2nd Int. Symp. Wave & Tidal Energy* no pp. nb.

[28] Anonymous, 1985, *Mersey Barrage, exploratory model tests*: London, Hydraulics Research Ltd; Carr, G.R., 1986, Studies of a tidal power barrage on the river Mersey: *Proc. 3rd Int. Wave, Tidal... Energy:* Paper No 31, 1–25.

attained by channeling it through ducts. Grant has discussed the potential use of the tidal flow for navigation buoys.[29]

So-called dynamic dams have also been proposed for tidal streams.[30]

## 8.7 Current Developments

The various turbine rotor options are, as has been said for some time, quite similar to those for wind turbines, the horizontal axial-flow turbine and the Darrieus or cross-flow turbine. In the latter type, blades rotate perpendicularly to the flow.

Options to secure a rotor include mounting the unit beneath a floating pontoon or buoy, suspending it from a tension leg arrangement between an anchor on a seabed and a flotation unit on the surface (as has been proposed in the past), or seabed mounting, easy only in shallow environments.[31]

### 8.7.1 Seaflow and Optcurrent

Canada is implementing a 250 kW demonstration plant[32] but Great Britain is installing and grid-connecting a 300 kW horizontal axis turbine. [33] The latter is a Joule Program project code named "Seaflow". The "Optcurrent" project is likewise a Joule Program undertaking involving Robert Gordon University and University College of Cork, besides IT Power.

The Seaflow project utilizes a Lynmouth turbine, a horizontal axis system mounted on a rigidly fixed vertical pillar. While the Stingray (see below) involves a linear lifted based device which relies on the same operational physical principles.

---

[29] Grant, A.D., 1981, Power generation from tidal flows for navigation buoys: *2nd Int. Symp. Wave, Tidal, OTEC.... Energy Proc.* 117–128.

[30] Yen, J.T. and Isaacs, J., 1978, Dynamic dam for harnessing ocean and river currents, and tidal power: *Proc. Ann. Mar. Technol. Soc./I.E.E.E. Comb. Conf. "Ocean '78"-"The Ocean Challenge"* **IV**, 582–584.

[31] Fraenkel, P., 2002, Energy from the oceans ready to go on-stream: *Renew. En. World* **8-2002**, 223–227.

[32] European Union (IT Power & Tecnomare), 2000, *Non-nuclear energy Joule II Project results:* Brussels, European Commission DG Sci (Rep. EUR-15683-EN).

[33] European Union, 1996–2002, World's first grid-connected tidal current turbine: Kassel, University of Kassel ISET.

## 8.7.2 Stingray

The "Stingray" project is underway (Fig.4). A feasibility study had been started in August 2001 and a prototype generator has been immersed off the Shetland Islands in Yell Sound, in 36 m deep water. Costing close to $3 million (€ 2 million), the generator weighs 180 metric tons; the 150 kW device was financed in part by the British government's Department of Trade and Industry. It was assembled on-shore along the Tyne River.

Due to the current's predictability the electricity can be marketed under the existing pool management regime free of under- or over-provision risk.

Stingray had a predecessor, the AWCG, a tidal stream device that used the current flowing over hydroplanes to lift a chamber up, and let it down, at the surface of the water.[34] The air in the chamber was alternately drawn in and expelled through a generator driving rotating turbine affixed on top of the chamber. The oscillating hydroplane principle was retained, though the hydroplanes are mounted on a completely submerged structure. Tidal current action on the hydroplanes initiates the oscillating motion that directly operates hydraulic cylinders. The cylinders act on high-pressure hydraulic oil that drives the generator. Seabed positioning protects the device from storms and insures that it does not interfere with navigation. An environmental impact assessment was conducted.

The hydroplane is 15 m wide and is installed at 20 m above the sea bottom. The structure is 24 m high. A yaw mechanism keeps the hydroplanes aligned with the flow of water through ebb and flow. A peak hydraulic power of 250 kW was matched by a time average output of 90 kW in a $1\frac{1}{2}$ m/s measured current. A repeatable 45 kW output was attained in a 1.7 m/s current speed.

Cost-wise a price of 8–30 US cents is foreseen; technological improvements will probably lower the price of the kWh.

The machine is due to get improved hydroplane control and a new configuration. Retrieved in later 2002, the newer Stingray version was re-deployed during 2003 for a longer period of operation. A correct grasp of the resource and the effects of placing multiple devices in service should be known later. According to the company involved—Engineering Business Ltd—a simultaneous program to start installation of a 5 MW version, connected with local power grids, was scheduled for July 2004.

Still in the Shetlands area, Scotland benefited from the European Union Regional and Urban Energy Programme.[35] Small islands would be happy recipients of electricity generated from ocean sources, e.g. Vlieland (see further below). In this project actual measurements were computer fed and, for two sites, the ensuing mathematical model showed that load factors of around 50% could be reached for 15 m–200 kW turbines rated for tidal current speeds of 2 m/s. Including installation

---

[34] ACWG = Active Column Water Generator.
[35] European Commission/Commission Européenne—Direction Générale XVII, 1995, *Feasibility study of tidal cureent power generation for coastal waters, Orkney and Shetland—Dossier 4/1040/92-41* : Brussels, European Union—European Commission.

and grid connection costs would run close to 920,000 Euros ($1.380.000[36]) for turbines with a minimum life-span of 15 years; electricity could be produced at a price of 0.123 Euros ($0.144) per kWh. With theoretically eight turbines with 20 m diameters a price of less than 0.74 Euros ($0.87) kWh would be reached.

### 8.7.3 Vlieland and the Electricité de France

The French electricity provider company, Electricité de France, through an intermediate subsidiary, has proposed to the island of Vlieland, in the Dutch province of Friesland, to be the location for one of its demonstrations of renewable energy projects. Among the technologies is a 2 × 2 tidal current turbine, consisting of two pillars each with two rotor blades. The pillars would be anchored to the seabed. The blades, 15 m wide, would rotate 18 h a day generating a total of 1,400 kW. Coastal protection would be a collateral benefit.

The project itself would combine tidal current, wind turbines, and hydrogen and fuel cells. Electrolytic hydrogen would be produced using surplus energy, a way to store power or to load fuel cells for public transport.

If the plans materialize, Vlieland would become the first island in the world whereon all power would be generated by and stored using exclusively renewable sources.

### 8.7.4 In the Arctic

The world's most northerly town, Hammerfest, in Norway, will be the first city to obtain its electrical power from a submarine station run by tidal currents. The 200 metric ton turbine is anchored on the seabed near Kvalsund. Its current capacity is 3 MW but it is to expand to 20 MW 36. The production would suffice to supply the needs of 1,000 homes. Costs have already reached $6.7 million ($\in$5,73 million) and by entire project completion should have had a price tag of $14 million ($\in$11.97 million). The cost of produced electricity at $0.04–0.05 ($\in$0.034–0.043) is however triple that of hydro-plant produced power in Norway.

This tidal power will be integrated to the electricity mix in the local grid. The turbine is similar to a wind turbine. The current speed is $2^1/_2$ m/s.

A risk factor is involved as storms have wrecked ocean power stations before. Success of the undertaking could transform Hammerfest in a tidal turbine manufacturing center. Studies have been conducted for Garolim, Korea and the Messina Strait in Italy. In Canada and Russia the Darrieus type turbine have had the favor for some time. It remains nevertheless that often plans have been laid to rest because of a major drawback, the low-energy density. All things being equal the energy from

---

[36] Exchange rate of May 2003.

a tidal current is one or two orders of magnitude lower than from a same diameter turbine in a barrage. It is felt that that disadvantage wipes out the savings provided by the dispensing with engineering works.

The authors do not stand alone lamenting that objections are continuously found to slow development of alternate ways to produce electrical power.

Commercial exploitation will have to solve the problems derived from waters either too deep or too shallow.[37] The technology to be developed is likely to be based upon buoyant tethered systems and not fixed seabed approaches. [38]

---

[37] MacNaughton, D. et al., 1993, Tidal stream turbine development. *IEE Conf. No 385(London) Abstr.*

[38] Bryden, I.; 2002, The future of tidal current power: *Oceans of Change Conf.—UK National Maritime Museum (30 October–November 1, 2002) Abstr.:* 4–5.

# Chapter 9
# Environment and Economics

## 9.1 Tidal Power and the Environment

But for the in depth environmental assessment conducted by T. Shaw and his collaborators, little or no environmental study has been carried out prior to the construction of Rance or Kislaya plants.[1] *Ex post facto* examinations were conducted on both schemes. At the Rance site the only milieu considerations were to protect the plant, and to insure a minimum of perturbations for the entrance to the port of St Malo. In fact the major changes have been the loss of *lanzons (Ammodytidae)*, the disappearance of some species and the take-over by others. High speed currents have been created near sluices and powerhouse, surges occur occasionally suddenly, sandbanks have disappeared, range of tides has decreased. Pool levels and navigation patterns have changed. Overall it appears that the TPP has affected the environment less than hydro-electric plants do.

Fisheries, an important aspect of St Malo-Dinard and surroundings economic life, have not been affected, industrialization has moved in, population has increased, tourism has not decreased. One can expect that no plant will ever be built again without a comprehensive assessment. Such a study would at least encompass following topics: water quality, wading birds, migratory fish, fish species elimination, development of intertidal algae, navigation, sediment movements, mudflats-marshes and sandbanks modifications, stratification increase, changes of pool water level, tidal regime modification[s].

In the case of the Bay of Fundy, shad has been slightly affected—notwithstanding the fish-path-way included in the plant—speed of current and range of tides somewhat reduced. Marsh drainage has been slightly unfavorably affected but storm damage reduced. A pre-construction study held that impacts would be kept at a minimum. Several impacts of biological nature would be eliminated if a scheme would not include a barrage.

As no barrage is needed, the plant requires less capital investment. A "removable barrage" scheme, which involves also re-timing has also been designed by Gorlov (1979, 1980). He proposes to use compressed air to insure re-timing; electricity can

---

[1] Shaw, T.L., 1979, *Environmental effects study of a Severn Estuary tidal power station*: Strathclyde, The University.

be generated directly or compressed air can be stored for later power production. The usual dam is to be replaced by a thin plastic barrier hermetically anchored to bottom and bay sides. A cable stretched across the bay or inlet, attached to several floats, would support the plastic barrier and hold it above water level. Environmental impact would be benign, though a pool of stagnant water may be created behind the barrier; pulled to one side, the barrier would allow evacuation of the impounded water and even navigation.

Most sites suitable for the construction of a tidal power plant are also locales of complex ecosystems, among others wintering and migratory birds. Intertidal flats which they use as feeding-grounds can be submerged by barrage-impounded water, and salt-marshes flora may be lost. Assuming that land pollution sources impact upon an estuary, the latter's impoundment may result in deterioration of water quality. Evidently effluent treatment before and after construction of the barrage may represent a non-entirely negligible additional cost, as tidal currents will be affected. Relief may however come through the presence of a larger mean volume of water (Parker 1993).

Lesser impacts, more site-specific, are related to coastal protection, flood control and groundwater. As co-lateral uses are looked for so as to lower the capital investment, or provide immediate additional returns, tourism development may be considered but is not necessarily in harmony with nature conservation (Table 9.1).

Assessments of environmental implications, *ex post facto*, have been numerous for the Rance River plants (e.g. Parker 1993, Retiere 1994). At least thirteen theses have been presented on the topic of environmental equilibrium in the area (Clavier 1981, Dauvin 1984, Grall 1972, Han 1982, Lacalvez 1986, Lang 1986, Lechapt 1986, Priou 1947, Retiere 1979, Rivain 1983). In fact the placing of cofferdams during construction, which cut the estuary, was the most damaging, a fact not reported in the Electricité de France's *post facto* assessments. Once in the operation stage things changed and an increasingly diverse fauna and flora colonized the area, showing a variable degree of biological adjustment; the ecosystems remain nevertheless strongly dependent on the operating conditions of the power station.

An environmental assessment was also conducted years after the Kisgalobskaia plant was constructed (1963–1968) and had operated for some twenty years. A first observation is that the water exchange between the bay and the open sea has been considerably reduced, salinity dropped in the upper fifteen meters of the water column while hydrogen sulfide accumulated below that level.

Additionally to those chemical modifications, the pre-construction bio-ecosystem seems to have been totally destroyed. When in 1987 water exchange was brought up to 30–40%, it triggered a slow-paced restoration of the ecosystem; by 1992–1993 fauna and flora did not differ substantially in distribution in and out the bay. The study conducted by Marfenin et al. (1997) reveals that the construction of tidal power centrals may have impacts which are far from benign and that a pre-construction impact study is absolutely necessary, at least where small plants in narrow entrance bays are concerned. The difference in parasite fauna in cod and pollock were studied for Kislaya and Ura inlets near Murmansk (Karasev et al.

9.1 Tidal Power and the Environment

**Table 9.1** Environmental impact assessments

| Author | Publication year | Site |
|---|---|---|
| Saleqzzaman | 2001 | Australia |
| Pierce | 2005 | Canada |
| Retiere | 1997 | France |
| Vantroys | 1957 | France |
| Waller | 1970 | France |
| Koh | 1977 | Korea |
| Bernshtein | 1997 | Russia |
| Karasev | 1996 | Russia |
| Marfenin | 1997 | Russia |
| Nekrasov | 1997 | Russia |
| Semenov | 1997 | Russia |
| Serguey | 2005 | Russia |
| Streets | 2003 | Russia |
| Usachev | 2004 | Russia |
| Baker | 1982 | United Kingdom |
| Dadswell | 1994 | United Kingdom |
| Duffett | 1987 | United Kingdom |
| Greenberg | 1982 | United Kingdom |
| King | 1977 | United Kingdom |
| Kirby | 1997 | United Kingdom |
| Longhurst | 1977 | United Kingdom |
| Little | 1977 | United Kingdom |
| Miles | 1977 | United Kingdom |
| Miles | 1981 | United Kingdom |
| Millichamp | 1977 | United Kingdom |
| Mitchell | 1981 | United Kingdom |
| Owen | 1977 | United Kingdom |
| Parker | 1993 | United Kingdom |
| Pratte | 1982 | United Kingdom |
| Shaw | 1975 | United Kingdom |
| Shaw | 1977 | United Kingdom |
| Tinkler | 1977 | United Kingdom |
| Wheatley | 1977 | United Kingdom |
| Daborn | 1982 | USA & Canada |
| Brooks | 1992 | United States |
| Haswell | 1981 | United States |
| Hogans | 1985 | United States |
| Hogans | 1987 | United States |
| Holmes | 1955 | United States |
| Gordon | 1994 | USA & Canada |
| Greenberg | 1977 | United States |
| Gordon | 1994 | United States |
| Larsen | 1981 | United States |
| Parker | 1993 | United States |
| Searratt | 1982 | United States |
| Streets | 2003 | United States |
| Van Walsum | 2003 | United States |

1996). Ecosystem research was further conducted by Semenov in 1997. It would be quite interesting, for comparison purposes, to have results of environmental studies—providing some were undertaken—of small Chinese power plants.

Change in tidal amplitudes, phases spectral composition of sea-level oscillations, tidal currents parameters are but a few of the modifications observed in the sea area around the power plant. These and other phenomena, such as suspended matter transport and movements of bottom sediments, can be estimated by modeling of tidal characteristics prior to tidal power plant implantation. Examples have been furnished for large plants that could be possibly constructed in the White and Okhotsk seas (Nekrasov and Romanenkov 1997).

The inner Bay of Fundy's estuaries—like other large amplitude estuaries—are home to numerous migratory fishes, such as the endangered sturgeon, herring, shad, bass, salmon but also to squids, sharks, seals and whales. Studies have shown that fishways at the Annapolis-Royal (Nova Scotia) power plant are not very effective in sparing marine life and mortality reaches, depending on such factors as species, size and turbine operation, 20 to even 80%. Were marine currents harnessed in the open ocean the impact of power stations may be such that a very serious decline in biopopulations could ensue. Dadswell and Rulifson's (1994) study follows up earlier ones by G.L. Duffett (*Tidal Power* 1987 p. 101) and by W.E. Hogans on mortality of adult fishes (*American Scientist* 1985 and 1987).

Never to be discouraged, opponents of a Fundy power plant raise the spectre of major ecological changes—besides fisheries losses—were a large station implanted in the upper reaches of the Bay. Variations in high (drop) and low tide (rise) sea level would compress the intertidal zone. New water levels for salt marshes would modify these habitats and modify primary production, which, however, would increase. Such changes would affect abundance of intertidal invertebrates, fish and migratory shore-birds (Gordon 1994).

Modelisation of water circulation in Passamaquoddy Bay was carried out, with a model forced by tidal height variations at the oceanic boundary, fresh water runoff from rivers and parameterized fluxes of heat and momentum at the sea surface and sea bottom. The natural system shows strong tidal currents in channels and passes, but near zero residual flow in the bay itself. If the tidal flood is reduced in the modified system, a significant tidal-residual flow passes from Passamaquoddy into Cobscook bay; freezing at the surface is more common with a power plant, but mainly due to a lower heat flux from the bay bottom. It thus looks as if some environmental impacts that had been mentioned as deterrents for a plant construction may well be less important than stated in the past (Brooks 1992, Holmes 1955). Tidal residual currents have also been modelized recently for the Juan de Fuca Strait and the Southern Strait of Georgia (Foreman et al.).

Retiere has examined repeatedly the environmental impact of the Rance River TPP and came to the conclusion, based on 20 years of operation, that the construction phase, during which the estuary was isolated, proved particularly environment damaging. However, once the plant was put into operation, a diverse flora and flora established itself. Their grouping into ecological units and their interrelationships, show a varied degree of biological adjustment. The operating conditions of the plant influences the fragile new ecological equilibrium, even though migratory organisms are able to pass through sluices and turbines. Little quantitative data is available that covers the pre- barrage situation. The comparisons of species distribution are thus based upon the known penetration into the estuary when there was no barrage. (Little & Mettam). Though these authors do not share the viewpoint that the reason for the small number of TPP is to be attributed to environmental concerns,—in his view the reasons are predominantly economic—one will agree that EIAs are useful tools to identify the impacts, and indeed applications of modern appropriate technologies might help abate the objectionable effects of a tidal power plant (Salequzzaman).

The environmental benefits of harnessing tidal power, among such other sources as hydro-electricity and extracting natural gas in the Russian Federation's Far Eastern region, were pointed out by Streets (2003). Neighboring areas are "choking" in the fumes emanating from coal-fired and bio-fueled electricity plants; cities in China, Mongolia, both Koreas and Japan would find present relief, not even mentioning considerable future benefits.

If Gibrat proposed a scale, or better formula, to rate the suitability of sites for location of a TPP, Sergey (2003) established a semi-quantitative rating of environmental disturbance brought about by power generating schemes. The thermal power industry is, according to that scale, the most damaging (coefficient 74) and the wave energy capturing industry the most benign (coefficient 31). Second to best is tidal power with a score of 42, just one point "better" than that of solar power.

Environmental and ecological assessments have been conducted in many instances and for different sites, thus sometimes before sometimes after construction of a TPP. Many of these "studies" are listed in this book's bibliography. Table 9.1 refers the reader to the "General Bibliography".

The project also named Roosevelt Island Tidal Energy Project (RITE) includes an impact study covering fish movement and concomitant protection of the biological resource (primordial concern), river navigation potential encumbrance and security, effect—if any—on the recreational and historical resources, and water quality. At the tip of the turbine's blunt rotor with a speed of 7.6 m/s there should be no impact upon fish. Further away from the rotor the speed is still lower. Furthermore turbines are spaced 12–30 m apart.[2] A protected area is foreseen around the system that covers in total $5,700 m^2$.

## 9.2 Economics

No major new technologies are needed for the current construction of tidal power plants, however, it may pay off to foster development and research the interface of a central's output with national grids, calculate a sound estimate of its economic

---

[2] 12 m east to west, three rows of two turbines each, 30 m apart; water depth 9 m.

interest, design and site proper implanting, and of course environmental effect and sustainability (Frau 1993).

The regularity from year to year of tidal power (less than 5% variation) is one of its main advantages. Co-lateral advantages are proper to a site: the dam can accommodate rail or road traffic, provide navigation improvements, cheap electricity a virtually inexhaustible supply of energy; it can constitute a send off for un- or poorly-exploited regions and is pollution free—though not entirely environmentally benign.

Though capital costs are high—but already reduced by one third by dispensing with cofferdams—the plant's useful life is two to three times longer than that of thermal or nuclear plants. Low-head water power could be converted to compressed air power, and smaller and cheaper high speed air turbo-generators would then produce electrical power.

As an example, for Bay of Fundy projects the benefits of a tidal plant due to fuel cost savings would exceed by far capital and operating costs; the best ratio of benefit to cost was estimated, ten years ago at 2.6 or 3. Compared to alternate energy sources, the benefits to the market areas for tidal power were found similar to nuclear, and superior to coal. These market areas cover the Maritimes, New England and New York; Quebec, a natural customer for such power, will not become a patron because it has too much hydropower available. The Canadian Board that conducted the study on the basis of un-re-timed output of single effect plants concluded that tidal power is economically viable. The simulation was carried out for a period spanning 1995–2015 with generation, loads and prices assumed to remain stable beyond 2015. The role of tidal energy would be to reduce the amounts needed of the most expensive fuel consumed in the market area.

Looked over in this estimation is the site of the plant, the huge compensation paid to farmers for their land and strikes that plagued construction.

The tidal power plant has no related fuel cost and thus once capital cost is recovered, cheaper electricity becomes available. A break-even level is reached at the time a same age nuclear facility or thermal plant has to be replaced. The tidal power plant offers other "dividends" because, contrary to coal-fired plants, it is free from sulfur dioxide and carbon dioxide pollution, acid rain generation, water pollution, oil spills, waste products treatment, and decommissioning expenses.

Development of low-pressure air turbine technology, on an industrial scale, was urged already some time ago to strengthen hydro-pneumatic power plants' competitiveness (Baker and Wishart 1986, Cave and Evans 1986).

The electricity production costs calculations sometimes favor the tidal plant, or are said to be about the same as those of conventional or nuclear plants, while the conventional stations are $CO_2$-polluters. Capital costs remain high for a tidal power plant but the longevity of the tidal plant is given at 75 years compared to 25 for a fossil fuel thermal central and between 30 and perhaps 40 years for a nuclear one.

Economic evaluation is made for construction costs per kW and generation costs, this being annual cost/annual electricity generated. The latter can be reduced by

## 9.2 Economics

increasing annual electricity generated and/or by reducing costs of equipment, dam construction (not applicable for tide current schemes).

If E represents the annual electricity generated, expressed in GWh, A is the surface area of the basin expressed in km$^2$, H the tidal range in m, and K a coefficient varying between 0.3 and 0.5, then

$$K = KAH^2$$

Annual costs are estimated to be 10–15% of the costs of total construction. No other costs are to be added in the ase of tidal power, but for thermal power, the costs of fuel (coal, oil, etc) must be.

Generation costs vary considerably from one country to another in function of a.o. social conditions; comparisons are thence all but meaningless. In Japan oil, gas and nuclear stations produce electricity at the lowest price with coal costing slightly more; tidal power would cost 3–4 times more, making it, currently, unattractive. Critics of tidal electricity generation labeled the Canadian Annapolis-Royal "an expensive undertaking".[3]

---

[3] The general topic of economics has been discussed in greater detail in Charlier (1982, *Tidal Energy*, op.cit.).

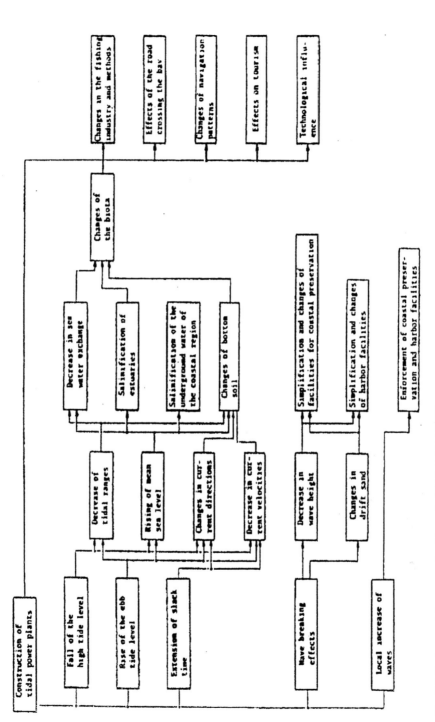

Fig. 9.1 Environmental assessment and impact of tidal power projects

# Annexes

Besides an "index", annexes include a comprehensive bibliography, a list of companies, research institutions, universities that are involved in matters relevant or pertaining to tidal power plants and related matters. It may be safely estimated that compared to the list compiled three years ago, there is an increase of approximately 30%.

The bibliography has a majority of works written in the English language, although considerable effort was expended to list French language contributions to the topic. [1] It has been arbitrarily cut off in 1982 and 1992. Papers, books and all other printed material pre-dating 1982 constitute the first section. The next one spans the 1982–1992 decade. If a motive for the 1982 cut-off has to be given, then let it be that it is the date when the first "modern" books dedicated to tidal energy/power were placed on the market (Charlier, [2] Baker), the fifteenth anniversary of the Rance River Plant was about to be celebrated, and some improvements in construction and operation had gained solid footing. This is not to say that some significant works dealing with TPPs had not been published, including *i.a.* those of Gibrat, Bernshtein, *et al.* before 1982.

The third section groups works placed into circulation since 1992. It might be appropriately pointed out that not less than 22% of all the publications fall within the 1992–2007 15 years span. It is the only one that includes, albeit in small number, publication not in "hard copy", *viz.* in printed form but also in electronic form (CDs, etc). This section is also less comprehensive because several authors list their own sources—works they themselves consulted—but without complete bibliographic details. The listings go to early 2007.

The General Bibliography encompasses most publications mentioned in footnotes. One exception to this policy concerns Chapter 2, wherein tide mills are discussed. A succinct bibliography has been appended at the end of that chapter to facilitate research on that specific subject.

---

[1] Some articles and books in German, Spanish, Portuguese, Russian or Dutch are similarly listed. Not to overlook Galician, Flemish, Catalan, etc. Where Chinese and Japanese language works are concerned, the titles have mostly been translated in English and a very few have had the original language title transliterated.

[2] See General Bibliography for the works mentioned herein.

Finally an intentionally kept-short glossary has been made into an annex. It is presumed that readers are today far more familiar with *ad hoc* terminology than they were in 1982 or 1993, dates of Charlier's previous books on this subject.

# Annex I: General Bibliography

## What was said before 1982

Abell, R., 1944. Tidal Power: *Overseas Engineer*, 12, 201, 204–206.
Ailleret, P., 1966. The place of tides in the development of the concepts of power generation, *Rev. Franc,. de l'Energie* **XVII**, 183, 642–659.
Alaska Power Survey, 1966, *Report of the hydro resources subcommittee of the Alaska Advisory Committee of the Federal Power Commission*, Juneau, State of Alaska.
All Chinese Conference on Tidal Power Utilization, 1959, *Proceedings*, Skuili-Yu-Tien-Li, Pamphlet Series. Shanghai, P.R.C.
Allard, P., 1951, Forme et énergie de l'onde-marée de vive-eau entre les eaux de Bréhat et le Cap de la Hague: *Annales de Géophysique*, vol. 7.
Allard, P., 1953, Influence de configuration des bassins sur le régime des marées littorales: *Annales de l'Institut Océanographique*, 28, 2, 63–112.
Allary, R., 1966, L'usine marémotrice de la Rance: La technique des travaux. *IV es J. de l'Hydraul. – La Houille Blanche* **XIX**.
André, H., 1976, Operating experience with bulb units at the Rance tidal power plant and other French hydro-power sites: *IEEE*, CH 1102-3-PWR.
Anderson-Nichols & Co., 1980, *Conceptual design of tidal air chamber. Report to Northeastern Univ.* The Company.
André, H., 1978. Ten years of experience at the "La Rance" tidal power plant, *Oc. Manag.* 4, 2–4, 165–178.
Andrews, J. H., 1977. Impact on wading birds. In Shaw, T. L. (ed.), *An environmental appraisal of the Severn Barrage*. Bristol, The University, pp. 100–106.
Anonymous, 1969, *The Rance tidal power plant* (in Japanese): Tokyo, Overseas Electrical Industry Survey Institute, inc. (November release).
Anonymous, 1970, *Report on the investigation on new electricity generation systems* (in Japanese): Tokyo, New Electricity Generation Systems General Investigation Committee, (September Report).
Anonymous, 1973, Present situations of the fundamental structures in the exploitation of the ocean (in Japanese): *Journal of the Japanese Society of Soil Mechanics and Foundation Engineering* no volume number, no page numbers.
Anonymous, 1980, Soviets make big tidal power plants: *World Water* 8, 8–9.
Anonymous (Construction, Chinese People's Republic), 1958, Experience in the topographical survey for a TPP (Russian translation from the Chinese) VINITI 12126/9.
Anonymous: see also at end of list (#1) anonymous works classified by year of publication Anthony, R. J., 1979. The changing times on tidal power, *Envir. Sci. & Techn.*, 13, 530–532.
Arnaud, C., 1958. *Le monde a faim de kilowatts*. Paris: Del Duca.
Back, P. A. A., 1978. Hydroelectric power generation and pumped storage schemes utilizing the sea, *Oc. Manag.* 4, 2–4, 179–206.

Baird, H., 1982. Upgrading of low density hydraulic energy, making it more acceptable for putting through a conventional hydroturbine, PICIVTTP 8, 1–8.

Baldwin, C. J., Houser, H.G. and Smith, H.L., 1964, Peaking capacity for power pools—pumped storage or gas turbines: *Proceedings, American Power Conference*

Banal, M. and Bichon, A., 1981. Tidal energy in France. The Rance tidal power station. Some results after 15 years of operation, *2nd Int. Symp. Wave and Tidal En.* (Cambridge), 327–338.

Barnier, L., 1968. Power from the tides, *Geographical Magazine* **50**, 1118–1125.

Barr, D. I. H., 1977. Power from the tides and waves. In Lenihan, J. (ed.), *Marine Environment* (vol. 5 of *Environment and man*). New York: Academic Press.

Barrett, M., 1981. Integrating tidal and wave power into the U.K. electricity systems, *Int. Symp. Wave and Tidal En.* (Cambridge).

Barton, R. (ed.), 1968. Hydrospace, *Quarterly Review of Ocean Management*. Bay of Fundy Tidal Power Review Board, Management Committee, 1976, *Preliminary report, Stage 1 of the Phase 1 Study Program*. Ottawa: Min. Mines

Behrman, S., and Thurlow, G. G., 1977. Use of colliery shale for construction. In Shaw, T. L. (ed.), *An environmental appraisal of the Severn Barrage*. Bristol, The University, pp. 132–135.

Bernstein, L. B., 1961a. *Central tidal-power stations in contemporary energy production*. Moscow: State Publishing House.

Bernstein, L. B., 1961b. "Tidal Power—A Russian view," *Canadian Consulting Engineer* (May).

Bernstein, L. B., 1961c. *Prilivniye elektrostantsu v sovremyennoy energetikye*. Moskva-Leningrad: Gosud energeticheskoye izdatyel'stvo.

Bernstein, L. B., 1964. The Rance River tidal-power plant (in Russian), *Gitsro-tekhnicheskoye Stroityel'stvo* **VI**, 46.

Bernstein, L. B., 1965a. *Tidal energy for electric power plants*. Jerusalem, Israel Program for Scientific Translation.

Bernshtein, L. B., 1965b. *Tidal energy for electric powerplants*. Translated from the Russian by Jerusalem, Israel Program for Scientific Translation, Springfield, Va., NTIS (378 p.).

Bernshtein, L. B., 1972. Kislaya Guba experimental tidal power plant and problems of the use of tidal energy. In Gray and Garhus (ed.), *Tidal power*: New York, Plenum pp. 215–238.

Bernshtein, L. B., 1974a. Russian tidal power station is precast offsite, floated into place, *Civ. Eng.* **44**, 4, 46–49.

Bernshtein, L. B., 1974b. Kislogubsk: A small station generating great expectations, *Water Power* **26**, 5, 172–177.

Berryman, M. S., 1979. Tidal energy and the energy crisis: an assessment of technology and the interrelationship. In *Marine Technology '79: Ocean Energy*. Washington: Marine Technology Society, pp. 107–116.

Bigourdan, G., 1920. Un moyen économique d'utiliser la force des marées, *Comptes-Rendus de l'Académie des Sciences (France)* **171**, 211–212.

Binnie, A. M., 1946, The effect of obstructions in tidal estuaries: *Engineering*, 161 (4183) 241–242.

Bird, E. C. F., 1978. Energy regimes and the Australian coast, *Ocean Management Conf.* **4**, 2–4.

Boardman, W. F., 1963. Quoddy and Rankin rapids as a multiple purpose project: the many surveys and reports have not yet produced a proper evaluation of Quoddy, or Quoddy–Rankin, as a multiple purpose project, *Public Utilities Forthnightly*, **71** (April 11), 19–25.

Boisnier, G., 1921a. *Utilisation de l'énergie des marées*. Paris: Annales des Ponts et Chaussées.

Boisnier, G., 1921b. *Utilisation de l'énergie des marées*. Rennes: Impr. Oberthur (96 p).

Bonnefille, R., 1963. Etude énergétique de la marée dans le golfe de St. Malo à partir des observations en nature, *Bulletin du Centre de Recherches des Etudes des Côtes*, **IV**, 153–165.

Bonnefille, R., 1976. Les réalisations de l'Electricité de France concernant l'énergie marémotrice: *La Houille Blanche* **31**, 2, 87–149.

Bonnefille, R., and Chabert–Dhières, G., 1967. Etude d'un modèle tournant de mer littorale. Application au modèle de l'usine marémotrice des îles Chausey, *La Houille Blanche* **XXII**, 6, 651–658.

Bonnefille, R., and Jeannel, M., 1964. Etude du modèle réduit de la coupure de la Rance, *La Houille Blanche* **XIX**, 4, 481–488.

Bonnefille, R. and Salomone, A., 1973, Comparaison de prototype et mesures de modèles sur échelle: *La Houille Blanche – Rev. Int. de l'Eau* 28, 2–3, 171–177.
Bourges, Y., 1966. The Rance tidal power scheme and the Saint Malo region, *Rev. Franç. de l'Energie.* **XVII**, 183, 861–863.
Bouteloup, J., 1950. *Vagues, marées, courants marins.* Paris: Presses Universitaires de France.
Braikevitch, M., 1972. Straight flow turbine. In Gray and Gashus (eds.) *Tida Power.* New York: Plenum, pp. 415–434.
Braikevitch, M. B., 1966. Contribution to the discussion on feasibility study of tidal power from Loughs Strangford and Caslingford with pumped storage at Rostrevor. *Proc. Inst. Civ. Eng.* 34, 1, 83–100
Brindze, R., 1968. *The rise and fall of the seas.* New York: Harcourt, Brace & World, pp. 72–82.
Butler, G., and Isen, H. C. K., 1966. *Corrosion and its prevention in waters.* New York: Van Nostrand Reinhold.
Cabanius, J., and Svilarich, E., 1966. The Rance project and its contribution to hydro-electric technology, *Rev. Franç. de l'Energie,* **XVII**, 183, 847–860.
Caillez, H., and Faral, M., 1966. General description of the tidal generating station and its electromechanical equipment, *Rev. Franç. de l'Energie,* **183**, 768–809.
Cattaneo, F., 1923. Rapport sur l'utilisation des marées, *Congrès Intern. de Navig.* (Landres) **XIII**, Sect. 2, *Comm.* 3.
Caquot, A., 1966. The definitive cut-off project, *Rev. Franç. de l'Energie,* **XVII**, 183, 712–721.
Caquot and Defour, 1937. *Utilisation perfectionnée de l'énergie des maréres.* Paris: Presses Universitaires de France.
Carlisle, N., 1967. *Riches of the sea: the new science of oceanology.* New York: Sterling Press, pp. 95–102.
Casacci, S. X., 1961, Advances in low-head machines, *Water Power,* **13**, 2, 62–67; **13**, 3, 104–108; and **13**, 4, 152–157.
Casacci, S. X., 1978a. Large bulb units for tidal power. Future prospects for electro-mechanical equipment: *Proc. Int. Conf. on Korea Tidal Power* (Seoul), 231–251.
Casacci, S.X., 1978b, Large bulb units for tidal power plants II : *Int. Water Power & Dam Constr.*[UK] 30, 7, 47–50.
Casacci, S.X., 1978c, Large bulb units for tidal power plants I: *Int. Water Power & Dam Constr. [UK]* 30, 6, 45–47.
Casacci, S. X., Duport, J. P., and Pariset, E. F., 1961. Research developments and results concerning bulb units: applications to river and tidal power plants, *Trans. Engineering Inst. Canada,* **IV**, 2.
Casacci, S. X., and Chapus, E. E., 1969. The bulb turbine, *Proc. Winter Meet. Assn. Mech. Eng.*
Casseau, M., 1962. Classification des cycles d'une usine marémotrice, *Mémoires et Travaux de la Société Hydrotechnique de France* **I** (supplement 76), 155–162.
Castro, J. F. M. de, 1956, Centrais Hidroeletricas utilisando o potencial das mares: Report of the 1st Congress of Engineers in Rio de Janeiro on the possibility of utilization of tidal power in Brazil, *Rev. Escola de Minas,* 20, 2, 23–27.
Cattaneo, F., n.d. *La transformazione della forza del mare in energia elettrica* (*Le forze motrici del mare*). Genoa: Stabilimento grafico editoriale [Piazza Saull, 5–2].
Center for Compliance Information, 1977. *The energy source book.* Aspen, Colorado: Aspen System's Corporation.
Charlier, R. H., 1968a. Tidal power, *Oceanology Intl.* **III**, 6, 32–35.
Charlier, R. H., 1968b. Marea ·i foamea de kilowatsi, *Progresele Stiintei* **IV**, 11, 481–485.
Charlier, R. H., 1977. Energy from the sea. In B. L. Gordon (ed.), *Marine Resource Readings.* Washington: University Press of America, pp. 115–161.
Charlier, R. H., 1969a. Tidal energy, *Sea Frontiers.* **XVI**, 6, 339–348.
Charlier, R. H., 1969b. La mer et la soif des kilowatts, *Revue de l'Université libre de Bruxelles* (Aug–Sept.), 17–33.
Charlier, R. H., 1969c. Harnessing the energies of the ocean, part I, *Marine Technology Soc. J.* **3**, 3, 13–22.
Charlier, R. H., 1969d. Harnessing the energies of the ocean, part II, *Marine Technology Soc. J.* **3**, 4, 59–81.

Charlier, R. H., 1970a. Harnessing the energies of the ocean—a postscript, *Marine Technology Soc. J.* 4, 2, 62–65.
Charlier, R. H., 1970b. French power from the English Channel, *Habitat* XIII, 4, 32–33.
Charlier, R. H., 1975. Comments to 'Power from the tides' by C. Lebarbier, *Naval Eng. J.* 87, 3, 58–59.
Charlier, R. H., 1978. Tidal power plants: sites, history and geographical locations, *Proc. Int. Symp. Wave & Tidal Energy (Canterbury)* I, 1, A1, 1–6.
Charlier, R. H., 1980. Tides and turbines, *Sea Frontiers* 26, 6, 355–362.
Charlier, R. H., 1981a. Tidal power, *Living Alternatives* II, 8, 31–34.
Charlier, R. H., 1981b. Energy from the ocean: a look at tidal power, *Alternative Sources of Energy* X, 50, 23–27.
Charlier, R. H., 1981c, Recursos energeticos del oceano: Ideas par un mundo mejor: *P.H.P Int* 2, 11, 4–19.
Charlier, R. H., 1981d, Whatever happened to tidal power: waiting for the political pull: *Oceans* 14, 6, 30–34.
Chen-Min-Chibao, 1958, Vzyat'energiyu ot morya (The seas as a source of power): Oct. 26.
Chi'u Hou Ts'ung, 1958. The building of the Shamen tidal power station: *Tien Chi-Ju Tung Hsin* 9, 52–56.
Chojniki, T., 1972. Calculs des marées terrestres théoriques et leur précision. *Polska Akad. Nauk. Zaklad. Geofizyki* 55, 3–42.
Civiak, R., 1978. Tidal power, *US Environmental Data and Information Serv.* 9, 5, 9–11 (Sept.).
Clare, R. and Oakley, A. J., 1981. The towing and positioning of caissons in a tidal barrage, *2nd Int. Symp. Wave and Tidal En.* (Cambridge), 177–190.
Clark, R. H., 1971. Recent tidal power investigations in the Bay of Fundy, Paper 2.2–55, *Proc. 8th World Energy Conf.* Bucharest, Romania (June. 27–July 2).
Clark, R. H., 1972a. Energy from Fundy tides. *Canadian Geographical J.* LXXXV, 5, 150–163.
Clark, R. H., 1972b. Fundy tidal power, *Energy International* 9, 11, 21–26 (Nov).
Clark, R. H., 1972c. La baie de Fundy, *2000-Grands Aménagements Mondiaux*, No. 83, 9–11.
Clark, R. H., 1973. La energia das marés na baia de Fundy, *Boletin Geografico*, No. 235 (Jul./Aug.).
Clark, R. H., 1976. Progress of reassessment of exploiting tidal energy, *E.I.C. Congress Proc.* 90.
Clark, R. H., 1978a. Reassessing the feasibility of Fundy tidal power, *Water Power and Dam Construction*.
Clark, R. H., 1978b. Power from the tides: *Geos.* Fall, 12–14.
Clark, R. H., 1978c. The economics of Fundy tidal power, *Proc. 1st Int. Symp. Wave & Tidal Energy (Canterbury)* I, 1, E3, 41–54.
Clark, R. H., 1979. Tidal power, *Proc. Int. Conf on Future Energy Concepts* [London 1/30–2/1, 1979].
Clark, R. H., 1980a. Organization and management for tidal power studies, *Proc. 1st Int. Symp. on Tidal Power*, Seoul, Korea (Nov. 14–15, 1978), Korean Inst. For Res. and Devel., pp. 105–110.
Clark, R. H., 1980b. Tidal power development, an international perspective, paper presented to *Int. Conf. on Performance of Concrete in a Marine Environment*, American Concrete Institute at St. Andrews-by-the-Sea (August 22, 1980), unpublished.
Clark, R. H., 1981a. Prospects for tidal power, *17th Session, Int. Conf. on Long-term Energy Resources*, UNITAR, Montreal (Nov. 26–Dec. 7, 1979).
Clark, R. H., 1981b. Fundy tidal power—a retrospective view, paper submitted December 1980, *International Journal of Ambient Energy*.
Clark, R. H., and Karas, A, N., 1979. Studies of tidal power from Bay of Fundy, *Int. Conf. on Future Energy Concepts, Institution of Electrical Engineers*, London (Jan. 30–Feb. 1, 1979), pp. 143–151.
Clark, R. H., and Walker, R. L., 1976. Progress of feasibility reassessment of exploiting tidal energy, *Proc. 90th EIC Congress*, Halifax, Nova Scotia (October).
Claude, G., 1921. Sur l'utilisation de l'énergie des marées. *Rev. Gén Electr.* 10, 627–631
Cochrane, S. R., and Wilson, E. M., 1981. The Stangford Lough tidal energy project, *2nd Int. Symp. Wave and Tidal En.* (Cambridge), 315–326.

Cohn, P. D., and Welch, J. R., 1969. Power sources. In *Handbook of Ocean and Underwater Engineering* (Myers, Holm, and McAllister, ed.). New York: McGraw-Hill, pp. 6–32 to 6–33.
Coleman, R. S., et al., 1976. Bibliography on pumped storage to 1975, *IEEE Trans., Power Apparatus and Systems* PAS-95, 3, 839–850.
Collins, J., 1977. Tidal energy, *LC Science Tracer Bulletin* TB 77-8. Washington Library of Congress, Science and Technology Division, (8 p.).
Collyns, G. S., 1951, The production of tidal power by the two-basin system: *Engineering*, 166, 392–294, 427–429.
Comyns, R. A., 1977. The use of China clay sand in construction. In Shaw, T. L., (ed.), *An environmental appraisal of the Severn Barrage*. Bristol, The University, pp. 146–150.
Considine, D. M. (ed.), 1977. *Energy technology handbook*. New York: McGraw-Hill, 1, 884 p.
Constans, J., 1978a. Present and future possibilities of energy production from mar-ine sources, *Proc. 1st Int. Symp. Wave & Tidal Energy (Canterbury)* I, 2, C1, 1–22.
Constans, J., 1978b. *Energy*. Monaco: Eurocean (Association Europée'enne Océanique), pp. 203–236.
Constans, J., 1979. *Marine sources of energy*. New York: Pergamon.
Cotillon, J., 1974. La Rance: Six years of operating a tidal power plant in France, *Water Power* 26, 10, 314–322. Ibidem, [In Serbian], 1977, *Elektroprivedra* 30, 5/6, 189–195.
Cotillon, J., 1978, Advantages of bulb units for low-head developments: *Int. Water Poiwer & Dam Constr.* 29, 1, 21–26.
Creager, W. P., and Justin, J. D. (eds.), 1950. *Hydroelectric handbook*, 2d Ed. New York: Wiley, 1, 151 p.
Creek, H., 1952, Tidal mill near Boston: *Civil Engineering*, 22, 840–841.
Crouzet, P. and Boissard, P., 1978, Pollution de l'estuaire de la Rance par les effluents de St Malo et Dinard [Pollution of the La Rance estuary by effluent from St Malo and Dinard]: *La Houille Blanche* 33, 7/8, 561–567.
Daborn, G.R., 1982, Environmental consequences of tidal power: *New Approach. Tidal Power Proc.*, Bedford Inst. Oceanogr., Dartmouth (N.S.), 11, 1–18.
Daborn, G. R. (ed.), 1977. *Fundy tidal power and the environment*. Wolfville, Nova Scotia: Acadia University Institute, 304 p.
Dalton, F. K., 1961. Tidal electric power generation, *J. Royal Astronomical Society of Canada*, 55, 1, 22–33; and 55, 2, 57–72.
Danel P., 1960. L'évolution de l'équipement des basses chutes, *Revue de l'Enseignement Supérieur* 1, 103–116.
Daric, G., 1957. Schéma de fonctionnement d'une centrale sous-marine équipression à fluide auxiliaire, *La Houille Blanche* (IV$^{es}$ Journées de l'Hydraulique) II, 694–701.
Davey, N., 1923. *Studies in tidal power*. London: Constable.
Da Vinci, L. (reprinted and edited by Dover Books in New York), *Notebooks*.
Davis, C. V., and Sorensen, E. E. (eds.), 1969. *Handbook of applied hydraulics*, 3d Ed. New York: McGraw-Hill.
de Bélidor, B. F., 1737. *Traité d'architecture hydraulique*. Paris: Ecole d'artillerie et du génie. Conservatoire National.
Debes, M., 1945. Utilisation de l'énergie des marées, *Revue de l'Institut Technique des Bâtiments et des Travaux Publics*, (no. d'avril).
Decelle, A., 1966. Twenty-five years of efforts, *Rev. Franc*$_2$. *de l'Energie*, **XVII**, 183, 636–641.
Decœur, 1890. *Appareil Hydraulique avec nouveau mode 'le de turbine pour l'utilisation continue de la force des marées:* Brevets Nos. 205–339 (29 avril 1890).
Deniaux, B., 1974. Tidal power generation in the French bay of Mont Saint-Michel, *Marine Affairs J.*, **2**, 97–115.
DePalma, J. R., 1968. An annotated bibliography of marine fouling for marine scientists and engineers, *J. Ocean Techn.* **II**, 4, 33–44.
De Rouville, A., 1957. General report on the utilisation of the tidal mechanical energy, *La Houille Blanche* **II**, 435–455.

Derrington, J., 1978a. The use of concrete caissons for river barrages, *Proc. Bristol Colston Symp.* (April).
Derrington, J. A., 1978b. Principles of design and construction for marine structures for wave/tidal/ocean thermal energy, *Proc. 1st Int. Symp. Wave & Tidal Energy (Canterbury)* I, 1, G2, 13–26.
Derrington, J. A., Turner, F. H., 1980. Floating concrete structures for energy production. *Proc. Concr. Ships and Floating Struct. Conv.*, Rotterdam, The Netherlands.
De Vos, F. J., 1957. Raisons pour lesquelles aucune usine marémotrice ne sera insérée dans le nouveau projet d'endiguement dit "Deltaplan," en Hollande, *La Houille Blanche* (IV$^{es}$ Journées de l'Hydraulique) II, 465–471.
Dhaille, R., 1957. Technique et rentabilité des dièdres à houle: *La Houille Blanche* (IV$^{es}$ Journées de l'Hydraulique) II, 421–429.
Dietz, R. S., 1972. Mineral resources and power. In Idyll, C. P. (ed.), *Exploring the ocean world: A history of oceanography*. New York: Crowell, pp. 164–195.
Dillon, G. S., 1975. Scope of tidal power from Indian Estuaries. *Indian J. Power River Der.* 8, 245–252.
Douma, A. and Stewart, G. D., 1981, Annapolis Straflo turbine will demonstrate Bay of Fundy tidal power concept: *Modern Power Systems* 1, 53–65.
Dubois, R., 1962. Les essais du groupe marémoteur expérimental de St. Malo, *La Houille Blanche* XV, 2, 131–140.
Duhoux, L., 1973, Comment se sont comportées les principales structures d' ingénierie civile [à la station marémotrice de la Rance]: *La Houille Blanche – Rev. Int. de l'Eau* 28, 2–3, 153–161.
Duron, G. S., 1975, Scope of tidal power from Indian estuaries, *Indian Journal Power and River Valley Development*, 245–252.
Dubrow, M. D., 1969. Tidal power. *Encyclopedia of the Earth Sciences*.
Duclos, M. I., 1958. Projets d'usines marémotrices en France. *Bull. Soc. Franc,.Electr.* 8, 85, 24–39.
Duhoux, L., 1964. Fermeture de la Rance; Déroulement des travaux et analyse des observations, *La Houille Blanche* XIX, 4, 491–508.
Easton, W. R., 1921. *Report on North Kimberley District*. Perth, Australia: North West Department Government.
Emery, K. D., 1974. Provinces of Promise, *Oceanus XVII*, 14–19.
Escande, L., 1967. Recherches hydrauliques récentes, *Arch. Hydrotechn. Polska*, **14**, 1, 3–13.
Fallon, P., 1964. The Rance estuary tidal power project, *Public Utilities Forthnightly* 70 (December).
Faral, M., 1973a. Les différents types de protection contre la corrosion mis en oeuvre à l'usine marémotrice de la Rance, *La Houille Blanche* 211, 2–3, 67–72.
Faral, M., 1973b, Corrosion-inhibiting methods at the Rance tidal power plant: *La Houille Blanche* 28, 2/3, 247–250.
Faral, M., 1973c, Méthodes anti-corrosives de la station marémotrice de la Rance: *La Houille Blanche – Rev. Int. de l'Eau* 28, 2–3, 247–250.
Fay, J. A. and Smachlo, M., 1981, Generic performance of small scale tidal power plants: *Proc. 2nd Int. Symp. Wave and Tidal Energy* (Cambridge) II, 409–419.
Fentzloff, H. E., 1972. The tidal power plant San José, Argentina. In: Gray and Gashus (eds.), *Tidal Power*. New York: Plenum.
Ferdinand, M., Hippert, M., Marolleau, Y., Dive. M., Morand. J. and Senellart, J. 1972, Wired and programmed automation applied to monitoring and operation of hydro-electric plants: *Automatisme* 17, 1–2, 3–21 [in French].
Ferdinand, M., Hippert, M., Marolleau, Y., Dive, M., Morand, J. and Snellaert, J., 1972, Automatisation relayée et programmée appliquée à l'observation et l'opération des stations hydroélectriques: *Automatisme* 17, 1/2, 3–21.
Fichot, E., 1923. *Les marées et leur utilisation industrielle*. Paris: Gauthier-Villars.
Fin-Chun-Yun, 1958, How to make the power of the oceans available to us; Report to the All-China Conf. on Tidal Power Utilization, *Shuilu-Yu-Tien Li*, no.2.

Firth, J. N. M., 1977. Pollution. In Shaw, T. L. (ed.), *An environmental appraisal of the Severn Barrage*. Bristol, The University, 93–99.

Fixel, A. E., 1969. Tidewater power generation, *Official Gazette U.S. Patent Office*, (February 11). **859**, 2, 415.

Fjelstad, J. E., 1965. Internal waves of tidal origin, *Geophysica Norvegica* **25**, 5, 1–73.

Fletcher, B. N., 1977. The geology of the area of the proposed barrage. In: Shaw, T. L., (ed.), *An environmental appraisal of the Severn Barrage*. Bristol, The University, 122–127.

Frieberger, A., and Collogen, C. P., 1964. "A laboratory methodology for studying marine fouling," *Feder. Soc. for Paint Techn.*, *Digest XXXVI*, **77**, 1198–1209.

Friedlander, G. D., 1964. The Quoddy question: time and tide, *IEEE Spectrum* **I**, 9, 96.

Furst, G. B., and Sud, S., 1977. Raw tidal energy absorption capability of a power system. *Proc. Summer Meet. IEEE (Mexico City)*, July.

Furst, G. B., and Swales, M. C., 1978. Review of optimization and economic evaluation of potential tidal power developments in the Bay of Fundy, *Proc. 1st Int. Symp. Wave & Tidal Energy (Canterbury)* **I**, 1, E2, 22–40.

Garrett, C., 1977. Predicting changes in tidal regime: the open boundary problem, *J. Phys. Oceanography* **7**, 2, 171–181.

Gandon, M., Guillaumin, M and de Larquier, M., 1973, Opération de l'usine marémotrice de la Rance: *La Houille Blanche – Rev. Int. de l'Eau* 28, 2–3, 131–144.

Grall, J. R., 1972, *Thèse* : Université de Paris.

Hermes, P., 1950a, Les projets d'aménagement de la Baie du Mont Saint-Michel pour l'utilisation de l'énergie des marées: *Le génie civil*, 6, 127, 16, 310–313.

Hermes, P., 1950b, Nouveau procédé de construction des digues pour usines marémotrices: *Génie Civil*, 127(17), 324–326.

Gauthier, M., 1962. A new approach to tidal power plant calculations, *La Houille Blanche* **XV**, 2, 259–275.

Gibrat, R., 1953. L'énergie des marées, *Bulletin Société Française d'Electricité* **VII**, 3, 283–332.

Gibrat, R., 1955. *Les usines marémotrices*. Paris: Electricité de France.

Gibrat, R., 1956. L'usine marémotrice de la Rance, *Revue Franç. de l'Energie* **VII** (avril).

Gibrat, R., 1957. Cycles d'utilisation de l'énergie marémotrice, Soc. Hydrotech. Fr. (IV cs J. de l'Hydraulique) *La Houille Blanche* **II** (Suppl.) 488–497.

Gibrat, R., 1962a. The first tidal power station in the world under construction by French industry on the Rance River, *French Technical Bulletin*, **2**, 1–11.

Gibrat, R., 1962b. Source de l'énergie des marées: énergie cinétique de la terre ou énergie thermique du soleil?, *La Houille Blanche* **XV**, 2, 255–266.

Gibrat, R., 1964. L'énergie des marées, *Rev. Française de l'Energie* XVII, 183, 660–684.

Gibrat, R., 1966a. *L'énergie des marées*. Paris: Presses Universitaires de France, 230 p. **XVII**, 183, 660–684.

Gibrat, R. V., 1966b. Energiya prilivov. V knigye: *Myezhdunarodnaya Assotsiatsiya po giravlicheskim issledovaniyan, XI Congress*, 1965, tom **6**, *Leningrad (St Petersburg)*, 223–242.

Gibrat, R., 1973. L'énergie marémotrice dans le monde: l'usine marémotrice de la Rance et l'environnement. *La Houille Blanche* **211**, 2–3, 145–151.

Gibrat, R., 1976. The current revival of tidal power studies. *Proc. A.S.T.E.D.*

Gibrat, R., and Auroy, F., 1956. Problèmes posés par l'utilisation de l'énergie des marées, *Fifth World Power Conference* (Vienna) **12**, 111, H/22, 4299–4328.

Gibson, A. H., 1933. Construction and operation of a tidal model on the Severn, *Appendix to the Report of the Severn Barrage Committee*. London: H.M. Stationery Office.

Gibson, H. C., 1966. The biological implications of the proposed barrages across Morecambe Bay and the Solway Firth, In Lowe-McConnell, R. H. (ed.), *Man-made lakes*. London: Academic Press.

Gibson, R. A., and Wilson, E. M., 1978a. Studies in retiming tidal energy. *Proc. Int. Symp. Wave et al.*: I, H–I, 1–10.

Gibson, R. A., and Wilson, E. M., 1978b. Tidal energy integration using pumped-storage, *ASCE. J. Power Div.*

Glasser, G., and Auroy, F., 1966. Research into the development of the bulb unit, *Rev. Franc,. de l'Energie*, **XVII**, 183, 722–767.
Godin, G., 1973. *The tidal power potential of Ungava Bay and its possible exploi-tation in conjunction with the local hydroelectric resources*. Canada, Marine Sci. Directorate, Manuscript Report Series 30.
Godin, G., 1972. *The analysis of tides*. Toronto: University of Toronto Press, 264 p.
Godin, G., 1974. The energetic resources of Ungara Bay and its hinterland. *IEE. Int. Conf.*, Halifax I, 378–383.
Gordon, F. R., 1964. *Secure Bay—Walcott Inlet tidal power scheme*. Gov. of West. Australia Preliminary geological report," Record No. 1964/6 (May 4).
Gorlov, A. M., 1978. Apparatus for harnessing tidal power: *US Patent* 4, 103, 490 (August).
Gorlov, A. M., 1979. Some new conceptions in the approach to harnessing tidal energy, *Proc. Miami Int. Conf. Alternat. Energy Sources* **II**, 1171–1795.
Gorlov, A. M., 1980a. A novel approach to exploitation of tidal power. Report: *US Dept. of Energy*.
Gorlov, A. M., 1980b. Small scale tidal energy consumption. In: Oktay, U. (ed.), *Energy resources and conservation related to the building environment*. London: Pergamon Press (pp. 492–498).
Gougenheim, A., 1967. The Rance tidal energy installation, *J. Inst. Nav.* **XX**, 3, 229–236 (July).
Gougenheim, A., 1976. L'utilisation de l'énergie des marées, *Cahiers océanographiques* **XIX**, 4, 277–293.
Gougenheim, A., and Romanovsky, V., 1957. Les remontées d'eau profonde, (IV$^{es}$ Journées de l'Hydraulique) *La Houille Blanche* **II**, 712–719.
Grant, A. D., 1981. Power generation from tidal flows for Navigation buoys, *2nd Int. Symp. Wave and Tidal En.* (Cambridge), 117–128.
Gray, T. J., and Gashus O. K., (eds.), 1972. *Tidal Power*. New York: Plenum, 630 p.
Great Britain House of Commons. Select Committee on Science and Technology, 1977. *Power in the Severn Estuary*. London: H.M. Stationery Office, 70 p.
Greenberg, D. A., 1976, Mathematical description of the Bay of Fundy numerical model: *Fish and Mar. Svc. Techn. Notes*.
Greenberg, D. A., 1977, Effects of tidal power development in the physical oceanography of the Bay of Fundy and the Gulf of Maine. In Daborn, G. R. (ed.), *Fundy tidal power and the environment*: Wolfville NS, Acadia University pp. 200–232.
Greenberg, D. A., 1979. A numerical model investigation of tidal phenomena in the Bay of Fundy and Gulf of Maine, *Marine Geodesy* **2**, 2, 167–187.
Griffin, D. M., 1974. *Energy from the ocean: an appraisal*. Washington, D.C.: Naval Res. Lab. (Memo. Rep. #2803), 43 p.
Guillaumin, M., and Larquier, M. de, 1973. Exploitation de l'usine de la Rance: méthode et résultats, *La Houille Blanche* **211**, 2–3, 131–144.
Halacy, D. S., Jr., 1977. *Earth, water, wind and sun. Our energy alternatives*. New York: Harper, pp. 59–71.
Han, S. J., 1983. Sedimentary conditions in tidal power plant areas of Garolim Bay, Korea and Rance, France. *Bull. Korea. Res. Dev. Inst.* 5, 1, 27, 34.
Harrah, B. K., and Harrah, D., 1975. *Alternate sources of energy*. Metuchen, NJ: Scarecrow Press, 201 p.
Hashimi, N. H., Nair, R. R., Kidwai, R. M., 1978. Sediments of the Gulf of Kachahh. A high energy tide dominated environment. *Indian J. Mar. Sci*.
Haswell, C. K., 1977. Civil engineering aspects. In Shaw, T. L. (ed.), *An environmental appraisal of the Severn Barrage*. Bristol, The University, pp. 128–131.
Haswell, C. K., Huntington, S. W., Shaw, T. L., Thorpe, G. R., and Westwood, I. J., 1972. Pumped storage and tidal power in energy systems, *Proc. Am. Soc. Civ. Eng.* PO 2, 201–220.
Heaps, N. S., 1968. Estimated effects of a barrage on tides in the Bristol Channel, *Proc. Inst. Civ. Eng.* **40**, 495–509.
Heaps, N. S., 1972. Tidal effects due to power generation in the Bristol Channel. In Gray and Gashus (eds.), *Tidal Power*. New York: Plenum, pp. 435–456.

Headland, H., 1949. Tidal power and the Severn Barrage, *Inst. Elec. Eng., Proc.* **96**, II, 51, 427–451.
Headland, H., 1950. Tidal power and the Severn Barrage, *Inst. Elec. Eng., Proc.* **97**, II.
Headland, H., 1951. Tidal power and the Severn Barrage, *Inst. Elec. Eng., Proc.* **98**, I.
Hermès, P., 1950, Les projets d'amée'nagement de la Baie du Mont St Michel pour l'utilisation de l'énergie des marées: *Le Génie Civil* 127, 16, 310–313.
Heronemus, H. E., Mangarella, P. A., McPherson, R. A., and Ewing, D. L., 1974. On the extraction of kinetic energy from oceanic and tidal river currents. In Stewart, J. B., Jr. (ed.), *Proc. MacArthur workshop on the feasibility of extracting useable energy of the Florida current.* NOAA Atlantic Oceanographic and Meteorological Laboratories (Miami).
Hiao Jing, 1957. Tidal power drive in turbo-pumps. *Chunkuo Shuili*, 9, 27.
Hooker, A. V., 1970. Severnside of the future. *Proc. Inst. Civ. Eng.* 47, 337–348; 49, 467–486.
Hughes, E. M., and Glanville, R., 1974. Tidal power, *Proc. Internat. Symp. On Altern. Energy Sources Centr. Elect. Bd.* 36–45.
Huntsman, A. G., 1928. *The Passamaquoddy Bay power project and its effect on the fisheries.* Saint John (New Brunswick), The Telegraph Journal.
International Passamaquoddy Engineering Board, 1959a. *Investigation of the international Passamaquoddy tidal power project: report to the International Joint Com-mission.* Washington, D.C.: The Board.
International Passamaquoddy Fisheries Board, 1959b. *Passamaquoddy fisheries investigations.* Report to the International Joint Commission. Washington/Ottawa: The Board, 53 p.
International Joint Commission, 1961a. Reports on Passamaquoddy Tidal Power Project, *U.S. Dept. of State Bull.* **44** (May 22).
International Joint Commission, 1961b. Rules against feasibility of Passamaquoddy project, *Electrical World*, **155**, (May 8), 44.
International Joint Commission, 1961c. *Investigation of the International Passama-quoddy Tidal Power Project. Report of the International Joint Commission, Docket 72;* Washington: The Commission (April).
Isaacs, J. D., and Seymour, R. J., 1973. The ocean as a power resource, *Int. J. Environ. Stud.* **3**, 4, 201–205.
Jefferys, E. R., 1981. Dynamic models of tidal estuaries, *2nd Int. Symp. Wave and Tidal En.* (Cambridge), 69–86.
Jeffreys, J., 1920. Tidal friction in shallow seas, *Philos. Trans.* **239**.
Jones, C., 1980. Quebec turns water into gold. *Christian Sci. Monitor* (July 30).
Jones, I. E., 1968. The Rance tidal power station, *Geography* **53**, 11, 412–415.
Kagan, B. A., 1974. Dissipation of tidal energy in the Arctic seas, *Akad. Nauk SSSR Bull. Atm. & Ocean. Phys. Ser.* **9**, 6, 375–376.
Kagan, B. A., 1977. Global dissipation and exchanges of energy between ocean and earth tides, *Akad. Nauk SSSR Bull. Atm. & Ocean. Phys. Ser.* **13**, 7, 485–490.
Kaiho, T., 1980, Self-stably moored and submerged water current powered rotor (in Japanese) *Resources* 207, 52–62.
Kammerlocher, L., 1957. Groupes générateurs hydroélectriques immergés, type bulbe. Développement et évolution constructive, *Rev. Générale de l'Electricité*, **66**, 7, 342–360.
Kammerlocher, L., 1958. Innovations technologiques dans la conception des groupes bulbes turbines–pompes immergés, *Soc. Hydrotechnique de France* (IV[es] Journées de l'Hydraulique) Aix-en-Provence, **V**, 2.
Kammerlocher, L., 1960. La station marémotrice expérimentale de Saint-Malo, *Rev. Génér. de l'Electricité*, **69**, 5, 237–261.
Karas, A. N., 1977. System planning for Bay of Fundy tidal power developments, *Proc. Summer Meet. IEEE (Mexico City)*, July.
Kay, R., 1975. Comments to "Power from the tides by C. Lebarbier," *Nav. Eng. J.* **87**, 3, 57–58.
Keiller, D. C. and Thompson, G., 1981. One dimensional modelling of tidal power schemes, *2nd Int. Symp. Wave and Tidal En.* (Cambridge), 19–32.

Kennedy, G. E., and Headland, H., 1957. Etudes de l'usine marémotrice de la Severn, *La Houille Blanche* (IV$^{es}$ Jour. de l'Hydraulique) **II**, 456–464.
Keough, M., 1959, Tidal power: *New Zealand Electrical Journal*, 32 (3), 82–83.
Keyerleber, K., 1973. Passamaquoddy: A good idea is hard to kill, *The Nation* **216**, 275–276.
Keyerleber, K., 1977. Tidal power: neglected energy resource, *Key Biscayne (Florida) Nat. Symp. on Energy and the Oceans: Conf. Coursebook* 7–58.
Kirby, R., and Parker, W. R., 1977. Sediment dynamics in the Severn Estuary: background for studies of the effects of a barrage. In Shaw, T. L. (ed.), *An environmental appraisal of the Severn Barrage*. Bristol, The University, pp. 41–52.
Koh, C. H., 1977, Korea megatidal environments and tidal power projects. Korean flats—Biology, ecology and land uses by reclamation and other feasibilities: *La Houille Blanche-Revue Int. de l' Eau* **52**, 3, 66–78.
Kohl, J. (ed.), 1976. *Energy from the oceans, fact or fantasy?* Conf. Proc., Jan. 27–28, 1976, Raleigh, N.C., Center for Marine and Coastal Studies, North Carolina State University. Sea Grant publication UNC-SG-76-04, 110 p.
[Korean] Ministry of Commerce and Industry, 1978, *Prefeasibility study of tidal power development* (Incheon Bay): Seoul, Res. Inst. of Shipping and Ocean.
Korea Electric Company, 1978, *Tidal power study 1978 – Phase I*: Seoul, Korean QC Res. & Dev. Inst. [August].
Kulev, I. P. et al., 1967, Electrochemical protection against corrosion of the structures of the Kislaya Guba tidal power plant: *Energet. Stroitel'stvo*, 4, 7.
Kuznetsov, E., 1979. *Analysis of a flexible barrier for a tidal power plant. A summary report to Northeastern University*. Columbus, OH: Battel Institute.
Laba, J. T., 1964. Potentials of tidal power on the North Atlantic coast in Canada and the U.S., *Coastal Eng.(Proc. 9th Conf. Coastal Eng.)* 832–857.
Laberge, N., 1976a, *Passamaquoddy tribe tidal project*: Passamaquoddy Tribal Council, Perry, Maine.
Laberge, N., 1976b, *Passamaquoddy Tribe tidal project*. Point Pleasant ME, Passam. Tribal Council.
Laberge, N., 1978. *Discussion papers of the Half-Moon Cove tidal power project*. Perry, Maine: Pleasant Point Reservation.
Lambert, M., and Legrand, M., 1973. Bilan de la protection cathodique à l'usine maréémotrice de la Rance, *La Houille Blanche* **211**, 2–3, 257–262.
Larsen, P. F., 1981, Some potential environmental consequences of tidal power development in the gulf of Maine and Bay of Fundy: (6th Biennal Int. Est. Res. Conf.) *Estuaries* 4, 3, 253–259.
Lawton, F. L., 1972a. The economics of tidal power. In: Gray and Gashus, *Tidal power*: New York, Plenum, 105–129.
Lawton, F. L., 1972b. Tidal power in the Bay of Fundy, In Gray and Gashus, *Tidal Power*. New York: Plenum, pp. 1–104.
Lawton, F. L., 1974. "Time and tide," *Oceanus* **XVII**, summer, 30–37.
Lebarbier, C. H., 1975a. Power from tides: The Rance tidal power station, *Nav. Eng. J.* **83**, 2, 57–71.
Lebarbier, C. H., 1975b. Power from the tides. Discussion, *Nav. Eng. J.* **87**, 3, 52–56.
Leborgne, M., Comportement des métaux à l'usine marémotrice de la Rance, *La Houille Blanche* **211**, 2–3, 251–256.
Lee, S. T., and Deschamps, C., 1977. Mathematical model for economic evaluation of tidal power, *Proc. Summer Meet. IEEE (Mexico City)*, July.
Lefrançois, J., 1973. Fonctionnement de l'usine de la Rance: comportement du matériel électromécanique, *La Houille Blanche* **211**, 2–3, 162–170.
Lefrançois, J. & Marolleau, Y., 1973, Operation of the Rance River tidal power plant and behaviour of its electro-mechanical equipment: *La Houille Blanche* 28, 2/3, 163–170.
Legendre, R., 1949. Les ressources énergétiques de la mer, *Bulletin Institut Océanographique Monaco* **947**, 1–16.

Legrand, J., 1973, Study of possible sea-water resistant materials: *La Houille Blanche – Rev. Int. de l'Eau* 28, 2–3, 263–269.

Legrand, R. and Lambert, M., 1973, L'expérience avec la protection cathodique à l'usine marémotrice de la Rance: *La Houille Blanche – Rev. Int. de l'Eau* 28, 2–3, 257–262.

Le Grand, R., and Lambert, M., 1962. Mesures électrochimiques appliquées à l'étude de la protection cathodique des ouvrages de la Rance, *La Houille Blanche* **XV**, 2, 177–186.

Le Grand, Y., 1957. Energie électromagnétique des océans, *La Houille Blanche* (Comptes-Rendu des IV[es] Journ. de l'Hydraulique, 1956) **I**, 225–228.

Leicester, R. J., Newman, V. G., and Wright, J. K., 1978. Renewable energy sources and storage, *Nature* **272**, 518–521.

LeMehaute, B., 1976. *A preliminary assessment of the tidal power potential at two sites in the vicinity of Cutler, Maine:* Arlington, Virginia: Tetra Tech, Inc., 43 p. (available NTIS AD-A023 824).

Lewis, J. G., 1963. The tidal power resources of the Kimberleys, *J. Inst. Eng., Australia*, **35**, 12, 333–345.

Licheron, S., 1962. La lutte contre la corrosion du matériel des usines marémotrices, *La Houille Blanche* **XV**, 2, 166–176.

Lingma, S. S., 1963, *Holland and the Delta Plan*: Van Nigh and Ditmar, Rotterdam.

Little, C., 1977. Possible biological effects. In Shaw, T. L. (ed.), *An environmental appraisal of the Severn Barrage:* Bristol, The University, pp. 61–71.

Logvenov, V., 1968. Prilivy stuzhat chelovyeku, *Pravda* 29, XII.

Longhurst, A. R., Radford, P. J., et al., 1977. Ecosystem models and the prediction of ecological effects. In Shaw, T. L. (ed.), *An environmental appraisal of the Severn Barrage*. Bristol, The University, pp. 83–92.

MacCellan, H. J., 1952. Energy consideration in the Bay of Fundy system, *J. Fisheries Board of Canada* **XV**, 2, 1935.

Macmillan, R. H., 1966. *Tides*. New York: Elsevier, pp. 172–179.

Magnien, M, 1977, Utilizing alternative energy in France: *Int. J. Energy Res.* 1, 1, 55–67.

Main, Ch. T., 1980, *Half Moon Cove tidal project:* Boston, Chas. T. Main Inc.

Maitland, W., 1782. History of London, London (Vol. I).

Malevanchik, B., and Natarius, R., 1968. *Luna v Zemnoy Upryazhkye*. Leningrad (St Petersburg), Gidrometeorologicheskoye Izdatyel'stvo.

Mappin, D., Gwynn, J. D., Hooker, A. V., et al., 1976. Proposals for a Severn Barrage, *Dock & Harbour Authority* **58**, 685, 333–338.

Marcellin, R., 1966. The Rance tidal power scheme, *Rev. Franç. de l'Energie*. **XVII**, 183, 628–631.

Mariano, 1438. *Utilization of tidal power* [in Latin]. Siena, Italy.

Mao, W. J., Beng, B. L., 1980. *An introduction to the development of small hydro-power in China*, New York, UNIDO.

Massé, P., 1966. The Rance tidal power scheme, *Rev. Franç. de l'Energie*, **XVII**, 183, 632–635.

Massé, P., and Gibrat, R., 1957. Application of linear programming to investments in the electric power industry, *Management Sci.* **I**.

Mauboussin, G., 1957. Construction de l'usine marémotrice de la Rance. Contribution des essais sur modèle réduit à la mise au point d'un mode d'exécution des travaux, *La Houille Blanche*. (IV[es] Journ. de l'Hydraulique), **II**, 388–399.

Mauboussin, G., 1966. Realisation of the Project, *Rev. Franç. de l'Energie*, **XVII**, 183, 810–846.

Maunsell and partners, 1976. *Kimberley tidal power study*. State Energy Commission of Western Australia.

May, T. P., and Weldon, B. A., 1954. Copper nickel alloys for service in sea water, *Intl. Congr. on Fouling and Marine Corrosion* (Cannes, France), June 8–13.

Maynard, 1919. Etude sur l'utilisation des marées pour la production de la force motrice, *Rev. Génér. d'Electricité* **22**.

McGown, L. B., and Bockris, J. O., 1980. *How to obtain abundant clean energy*. New York: Plenum.

McHugh, J. L., 1961. Power from the tides, *Sea Frontiers* **VII**, 1.
McKeough, 1959, Tidal power: *New Zealand Elect. J.* 32, 3, 82–83.
McNaughton, A. G., 1960. "Passamaquoddy not feasible?" *Electrical World*, **153** (March 28), 60–61.
McQueen, H. J., 1977. Critical analysis of changes in the Canadian energy indus-tries. In Wilmot, P.D., and Slingerland, A. (eds.), *Technology assessment and the oceans:* Boulder, Colorado: Westview Press.
Meerwarth, K., 1963. *Wasserkraftmaschinen, ein Einfuehrung in Wesen, Bau und Berechnung von Wasserkraftmashinen und Wasserkraftanlagen.* Berlin: Springer Verlag.
Miles, G. V., and Webb, D. G., 1977. Influence of Severn Barrage on the tidal regime. In Shaw, T. L. (ed.), *An environmental appraisal of the Severn Barrage.* Bristol, The University, pp. 35–40.
Miles, G. V. and Worthington, B. A., 1981. The influence of Severn tidal power schemes on sediment transport processes, *2nd Int. Symp. Wave and Tidal En.* (Cambridge), 105–116.
Miller, H., 1975. *Die Straflo Turbine, die technische Realisation von Harza's Ide"en.* Zurich: Straflo-Group.
Miller, H. and Wilson, E. M., 1981. Optimizing Straflo turbines for tidal applications, *2nd Int. Symp. Wave and Tidal En.* (Cambridge), H2.
Millichamp, R. I., and Staite, E. M., 1977. Fisheries considerations. In Shaw, T. L. (ed.), *An environmental appraisal of the Severn Barrage.* Bristol, The University, pp. 112–121.
Milne, A. J., 1978. Tidal power in Scottish sea lochs, *Proc. Int. Symp. Wave & Tidal Energy (Canterbury)* I, 2, C2, 23–24.
Fox, W., Brooks, B., and Tyrwhitt, J., 1976. *The Mill.* Toronto: McClelland & Stewart. [French] Ministère de l'Industrie et de la Recherche, 1976, L'usine marémotrice de Chausey. In *La Production d'électricité d'origine hydraulique; Dossier de l'Energie 9*, Chapitre IV, 41–49: Paris, La Documentation Française.
Mettam, C., 1977. Impact on the intertidal system. In Shaw, T. L. (ed.), *An environmental appraisal of the Severn Barrage.* Bristol, The University, pp. 72–82.
Mitchell, R. Probert, P. K., and McKirdy, A. P., 1981. Environmental implications of tidal power; proposals with particular reference to the Severn Estuary: *Proc 2nd Int. Symp. Wave and Tidal En.* (Cambridge), 2, paper D-2.
Monteiro, A. de F., 1953. *Le Problème de la Capture de la Rance;* Mémoires et Documents, Laboratoire de Géomorphologie (No. 5).
Monteiro de Castro, J. F., 1956, Centrais hidroelectricas utilisando o potential das mares: *Rev. Escolas de Minas* 20, 2, 23–27.
Moreau, G., 1931. *Etude sur l'utilisation de l'énergie des marées en France.* Paris: Delagrave.
Musgrove, P., 1979. Tidal and river current energy systems. *Proc. 1st Intl. Conf. Future Energy Concepts, Inst. Elec. Eng.(London)* (IEEE Conf. Public No. 17). pp. 114–117.
Nachamkin, M., 1979. *Comparative analysis of various prime movers for recovering of tidal energy. A summary report to Northeastern University.* New York: Gibbs and Hill.
Holmes, C., 1955, *Tidal power in Passamaquoddy Bay. A thesis:* Bangor, University of Maine. (Japanese) National Institute of Resources, 1979, *Report on the survey on the gradual utilization of the kinetic energy of the Kuroshio*: Tokyo, National Inst. of Resources (Report '82, in Japanese).
Nekrasov, A. M., and Posse, A. V., 1959. Work done in the Soviet Union on high-voltage long-distance direct current power transmission, *Trans. Austral. Inst. Elect. Eng.*, Part III-A: *Power Apparatus and System*, **78**, 515–521.
New England–New York Inter-Agency Committee, 1955. *Report:* New York, "The Committee" (See Part II, Chapter XI, "Special Subjects," Subregion A, Section II).
NOAA, 1978. *Harnessing tidal energy.* Washington: Department of Commerce.
Noyes, R. (ed.), 1977. *Offshore and underground power plants.* Park Ridge NJ: Noyes Data Corp.
Oh, S.W. and Van Walsum, E., 1979, Korea tidal power and beyond: *2nd Miami International Conference on Alternative Energy Sources* no pp. numbers.
Olivier-Martin, D., Lebailly, P., and Banal, M., 1966. The studies and methods for barring the Rance. *Rev. Franc. de l'Energie*, **XVII**, 183, 685–711.

Owen, M., 1977. Implications for windfowl. In Shaw, T. L. (ed.), *An environmental appraisal of the Severn Barrage.* Bristol, The University, pp. 107–111.
Parenty, H., and Vandamme, G., 1920. Utilisation de la force des marées et du choc des vagues de la mer. *Comptes-Rendus de l'Académie des Sciences* [Fr.] 171, 896–898.
Passamaquoddy-St. John, 1964a. *Hearing before a sub-committee of the Committee on Public Works United States Senate, 88th Congress. Second Session on S. 2573.* Washington, D.C.: U.S. Government Printing Office.
Passamaquoddy-St. John River Study Committee, 1964b. *Supplement to the July 1963 Report, The International Passamaquoddy Tidal Power Project and Upper Saint John River Hydro-electric Power Development.* Washington, D.C.: Department of the Interior.
Penner, S. S., and Icerman L. (eds.), 1974–6. "Tidal and wave energy utilization." In *Energy*, vol. 2. Reading, Mass.: Addison-Wesley.
Pogvinov, V., 1968. Prilivy stuzhat chelovyeku. *Pravda*, **29**, XII.
Poteryakhin, A., 1935, *GES na Prilivno-Otlionykh kolebaniyak v ust e r.Mezen* [Thesis]: Moscow, State Publishing House [I.S.I.].
Priou, M., 1947, *Thèse:* Université de Rennes.
Proctor, R., 1981. Mathematical modelling of tidal power schemes in the Bristol Channel, *2nd Int. Symp. Wave and Tidal En.* (Cambridge), 33–52.
Public Works Department, 1961. *Tidal predictions, North West Coast.* Perth: Western Australia, Harbors and Rivers Branch.
Rabaut, J., 1981, Description of "La Rance" tidal power plant: Detection, Diagnosis and Prognosis: Contribution to the Energy Challenge. *Proc. 32nd Meet. Mechan. Failures Prevention Gr.* 289–309.
Rath, R., and Surrel, G., 1957. La corrosion par la mer du matériel des usines marémotrices. *Mémoires & Travaux Soc. Hydrotech. France* **II**, 139–143.
Raynor, C. J., 1950. *Power from the tides with special reference to Western Australia* [Thesis]. Perth: University of Western Australia.
Remenieras, G., and Smagghe, P., 1957. Sur la possibilité d'utiliser l' énergie des courants marins au moyen de machines à aérogénérateurs, *La Houille Blanche* (IV$^{es}$ Journ.de l'Hydraulique) **II**, 532–539.
Renne, R. R., 1966. The future of water resources, *Oceanol. Int.* **I**, 2, 67–71.
Retiere, C., 1979, *Thèse:* Université de Rennes-1.
Richards, A. F., 1976. Extracting energy from the oceans. A review, *Marine Techn. Soc. J.* **10**, 2, 5–24.
Richards, B. D., 1948. Tidal power, its development and utilization, *J. Inst. Civ. Eng.(Great Britain)*, 104–144 (April).
Rigaud, M., 1926. *Les réserves d'énergie.* Paris: Gauthier-Villars.
Robinson, I. S., 1981. Seiching in tidal power basins: can it increase tidal output? *Proc. 2nd Int. Symp. Wave and Tidal En.* (Cambridge), 2, 53–68.
Romanovsky, V., 1950. *La mer, source d'énergie.* Paris: Presses Universitaires de France.
Rouch, J., 1961. *Les marées.* Paris: Payot, Bibliothèque Scientifique, 230 p.
Rouzé, M., 1959. *Energie des marées.* Monte-Carlo: Editions du Cap.
Ruxton, T. D., 1982. Overview of the environmental aspects of the Severn Development. *Int. Conf. New Appr. to tidal power III*, 3–5.
Ryan, P. R., 1979–80, Harnessing power from the tides: *Oceanus*, 22, 4, 64–67.
Saji, Y. Yoshiwa, M. and Iwata, A., 1980, Study of electromagnetic generation system of ocean energy extraction (in Japanese): *Resources* 207, 40–51.
Sandrin, P., 1980, AGRA: a new operational management model for La Rance tdal power plant: *Bull. Dir. Et. & Rech. Sr. B* [1980], 3, 29–40.
Sanhes, J., 1962. Protection contre la corrosion marine de la station marémotrice expérimentale de St. Malo, *La Houille Blanche* **XV**, 2, 195–204.
Saunders, D. W., 1975. Kimberley tidal power revisited, *J. Inst. Eng. Austral.* 47, 11, 47–55.
Saunders, D. W., 1976, Kimberley tidal power: *Proc. ANZAAS Congr.*

Savage, J. A., 1975. *Potential of tidal power and Gulf Stream power sources.* Austin: Governor's Energy Advisory Council, 49 p.
Savery, S. P. A., 1977. Natural aggregates supply considerations. In Shaw, T. L. (ed.), *An environmental appraisal of the Severn Barrage:* Bristol, The University, pp. 136–145.
Saylor, J. P., 1965, The Passamaquoddy boondoggle: economic feasibility of utilizing high tides near the Maine–New Brunswick border to generate electric power: *Public Utilities Fortnightly,* 71 (January 17), 15–22.
Schureman, P., 1975, *Tide and current glossary:* (Revised edition by Steacy-Hicks) Washington, D.C.: National Ocean Survey (25 p.)
Scott, W. E., 1976a. Australia takes new look at tidal energy, *Energy Intl.* **13**, 9, 41–43.
Scott, W. E., 1976b. Western Australia examines energy options: *Energy International,* **13**, 9, 25–28.
Sebo, S. A., 1975. Ocean powers, *Maritime Studies & Management* **2**, 4, 202–214.
Secretary of the Interior, 1964. *The international Passamaquoddy tidal power project and Upper Saint John River hydroelectric power development.* Washington, D.C.: Department of the Interior.
Seifert, A., 1948. Gezeitenkraftwerk in Wilhelmshafen, *Arch. Energiewirtschaft* **IV**, 209.
Select Committee on Science and Technology, Energy Resources Subcommittee, 1975. *Tidal power for electricity generation. A memorandum from the Central Electricity Board:* London: HMSO (Part IV).
Select Committee on Science and Technology, Energy Resources Subcommittee, 1977. *The exploitation of tidal power in the Severn estuary.* London: HMSO (4th rep.).
Seoni, R. M., 1977. Major electrical equipment proposed for tidal power plants and the Bay of Fundy, *Proc. Summer Meet. IEEE (Mexico City)*, July.
Severn, B., and Campbell, R. O., 1978. Prefabricated caissons for tidal power development, *Proc. 1st Int. Symp. Wave & Tidal Energy (Canterbury)* **I**, 1, G1, 1–12.
Severn Barrage Group, 1976. *Proposal for the project. Definitive study of the Severn Barrage.* London: David Mappin (Offshore) Ltd.
Shaw, T. L., 1974. Tidal energy from the Severn Estuary, *Nature* **249**, 5459, 730–733.
Shaw, T. L., 1975. Tidal power and the environment, *New Scientist* **68**, 972, 202–4/206.
Shaw, T. L., 1976, Tidal power closing the gap: *Water Pow. & Dam Constr.* (May) 24–25.
Shaw, T. L., 1977a. A policy for tidal energy, *Marine Policy* **1**, 61–69.
Shaw, T. L., 1977b. Tides, currents and waves, In Shaw, T. L. (ed.), *The exploitation of tidal power in the Severn Estuary.* Bristol, The University, pp. 1–34.
Shaw, T. L., 1978a. Tenth world energy conference, *Water Power & Dam Construction,* January, 58–62.
Shaw, T. L., 1978b. The status of tidal power, *Water Power & Dam Construction,* 30, 6, 29–25.
Shaw, T. L., 1978c. The role of tidal power stations in future scenarios for electricity storage in the U.K., *Proc. 1st Int. Symp. Wave & Tidal Energy (Canterbury)* June, 29–34 **I**, 1, H2, 11–22.
Shaw, T. L., and Thorpe, R. G., 1971. Integration of pumped storage with tidal power, *Proc. Am. Soc. Civ. Eng.*PO **1**, 159–180.
Shaw, T. L., and Westwood, I. J., 1974. Optimising pumped storage with tidal power in an estuary, *Amer. Soc. Mech. Eng.* Paper in 74-WA/pwr-7, 7 p.
Shaw, T. L., and Wheater, H. S., 1978. Some observations on the virtues of integrating tidal power in the U.K. electrical network. *Proc. Bristol Univ. Colston Res. Symp.*
Shepard, F., 1949. Evidence of world-wide submergence, *J. Mar. Res.* **VII**, 661–676.
Sibley, A. K., and McNiece, W. H., 1960. Harnessing the tides, *Military Engineer,* 52, January/February, 1–6.
Simonet et Jacquinet, 1954, Les usines marémotrices. Projet de la Baie du Mont St Michel. *Inforn. Ge'ogr.* 4, 147–148.
Skillman, J. M., and Wheater, H. S. 1977. The prospects of tidal power, *Inst. Civ. Eng., Proc.* (Part 1—Design and construction) **62**, 701–705.
Smith, L., 1959. The Quoddy project stirs again, *Public Utilities Fortnightly,* **64** (November 5), 753–765.

Smith, L., 1961. The status of power supply in Maine, *Public Utilities Fortnightly*, **68**, (December 7) 873–882.
Smith, P. C., 1978. Circulation, variability and dynamics of the Scotian shelf and slope, *Fisheries Research Bd. J.(Canada)* **35**, 8, 1067–1083.
Sogreah, 1959a. *Usina maremotriz del golfo de San José Anteproyecto.* Grenoble, France: Société d'Etudes et des Applications Hydrauliques.
Sogreah, 1959b. *Construcion del Canal San Jose'*. Grenoble, France: Société d'Etudes et des Applications Hydrauliques.
Sogreah, 1959c. *Anexo, Calculo aproximado de la influencia de la rugosidad de la inercia del agua sobre la caida turbinable.* Grenoble, France: Société d'Etudes et des Applications Hydrauliques.
Sogreah, 1963. *Tidal power plant for Collier Bay.* Grenoble, France: Société d'Etudes et des Applications Hydrauliques (Report No. R8527 for the Ministry of Industrial Development of the Government of Western Australia).
Sogreah, 1965. *Collier Bay tidal power development: Secure Bay.* (Report No. R9011 to the Public Works Department of the Government of Western Australia).
Song, M., 1981, China is interested in tidal energy: *Rev. de l'Energie* 32, 337, 438–439.
Song, W., 1979, Korea tidal power project Phase I study: *Proc. Int. Symp. Korea Tid. Pow.* (Nov, Seoul).
Sorokin, V. N., 1959, *Prilivnaya Elektrostantsiya v Limbovskom Zalive* [Thesis] Moscow, Energy Institute.
Stokes, C. J. and Street, R. D. J., 1981. Turbine caissons for the Severn barrage, *2nd Int. Symp. Wave and Tidal En.* (Cambridge), 167–176.
Subrahmanyam, K. S., 1978, Tidal power in India: *Water Power & Dam Constr.* 6, 42–44.
Swales, M. C., and Wilson, E. M., 1966. Optimisation of tidal power generation, *Water Power* **20**, 3, 109–114.
Tanner, R. G., 1979. Tidal energy in the Bay of Fundy. In *Marine Technology '79: ocean energy.* Washington, D.C.: Marine Technology Society, pp. 91–99.
Teng, A. A. et al., 1979, *The role of small hydropower generation in the energy mix development for the P.R.C.*: Palo Alto CA, Oriental Engineering and Supply Co.
Terry, R. D., 1966. *Ocean Engineering.* Energy sources and conversion, undersea construction, vol. 3. No. Hollywood, California: Western Periodicals.
Thompson, I. B., et al., 1967. *The St. Malo Region.* Nottingham (England) Geography Field Group. Regional Studies No. 12, pp. 8, 16, 19, 63, 69–72, 75–76, 95.
Tidal Power Review Board, 1977. *Reassessment of Fundy tidal power*: Ottawa, Bay of Fundy T.P. Rev. Bd., pp. 24–40.
Tinkler, J. A., 1977. Drainage and land quality. In Shaw, T. L. (ed.), *An environmental appraisal of the Severn Barrage.* Bristol, The University, pp. 53–60.
Trites, R. W., 1959. *Probable effects of proposed Passamaquoddy power project on oceanographic conditions.* Intl. Passamaquoddy Fisheries Bd. Rep. to Int. Joint Commission, Ch. 7, Appendix 1.
Tuthill, A. H., and Schillmoller, C. M., 1965. *Guidelines for selection of marine materials.* New York: International Nickel Company.
Tuttel, J., 1978. Watermolens eeuwenoud en eeuwig boeiend, *Aard & Kosmos* **8**, 9 en 10.
Udall, S. L., 1963, *The international Passamaquoddy tidal power project and Upper St John River hydroelectric power development*: Washington, DC, US Dept. of the Interior.
U.N. Department of Economic and Social Affairs, 1957. *New sources of energy and economic development.* New York: United Nations, 150 p.
U.S. Army Corps of Engineers, 1964. *Supplementary engineering report to the Report on the International Passamaquoddy Tidal Power Project and Upper Saint John River Hydroelectric Power Development:* Waltham, Mass.: U.S. Corps of Engineers, New England Division.
U.S. Army Corps of Engineers, 1980, *National Hydroelectric power study XV.*
U.S., Congress, House Committee on Foreign Affairs, 1953. *Survey of Passamaquoddy tidal power project.* Hearings before the subcommittee of the Committee on Foreign Affairs, House of

Representatives. H. J. Res. 112, H.J. Res. 113, and H. J. Res. 114 to authorize and direct the International Joint Commission on United States–Canadian Boundary Waters to make a survey of the proposed Passamaquoddy tidal power project and for other purposes. July 14 and July 22, 1953. Washington, D.C.: Government Printing Office (83rd Cong., 1st sess.).

U.S., Congress, 1955. *Passamaquoddy international tidal power project*. Report on S. J. Res. 12, a resolution requesting the Secretary of State to arrange for the International Joint Commission, United States and Canada, to conduct a survey of the proposed Passamaquoddy tidal power project. Washington, D.C.: Government Printing Office (84th Cong., 1st sess., H. Rept. 1152).

U.S., Congress House, 1965. *Communication from the President of the United States*, Lyndon B. Johnson. Conservation of the Natural Resources of New England. Report on the Passamaquoddy-St. John River Basin Power Development, together with a recommendation for the immediate authorization of the Dickey-Lincoln School Project on the St. John River. Washington, D.C.: Government Printing Office (89th Cong., 1st sess., H. Rept. 236).

U.S., Department of the Interior, 1961. *Report to Passamaquoddy-Saint John River Study Committee*. The International Passamaquoddy Tidal Power Project and Saint John River, United States and Canada, Load and Resources Study.

U.S., Department of the Interior, 1964. Office of the Secretary. Fact Sheet—August. (Release: Wood 343–3171.) *The Passamaquoddy international tidal power project and Upper Saint John River Hydroelectric Development.*

U.S., Energy Research and Development Administration, Technical Information Center, 1976. *Solar energy: A bibliography*. Oak Ridge, Tennessee, 2 vols.

U.S., Federal Power Commission, 1941. *Passamaquoddy tidal power project*. Letter from the chairman of the Federal Power Commission, transmitting in response to S. R. No. 62 (76th Cong.) a report on the Passamaquoddy tidal project, Maine. Washington, D.C.: Government Printing Office (76th Cong.).

U.S., Library of Congress. Congressional Research Service, Science Policy Research Division, 1978. *Energy from the ocean: Report prepared for the Subcommittee on Advanced Energy Technologies and Energy Conservation Research, Development and Demonstration of the Committee on Science and Technology*. Washington, D.C.: Government Printing Office (Chapter 4, Energy from ocean tides, pp.175–222).

U.S., Passamaquoddy-Saint John River Study Committee, 1963. *The International Passamaquoddy tidal power project and Upper Saint John River hydroelectric power development*. Report to President John F. Kennedy in response to letter of May 20, 1961, submitted by Stewart L. Udall, Secretary of the Interior. Washington, D.C.: Printing Office, 95 p.

U.S., Senate, Committee on Public Works, Subcommittee on Flood Control, Rivers and Harbors, 1964. Passamaquoddy-St. John appendix material compiled on conjunc-tion with hearing on S. 2573, *a bill to authorize the international Passamaquoddy tidal power project*. Committee Printing, 88th Cong., 2nd sess.

U.S., Senate, 1964. *Passamaquoddy-St. John Hearing*, August 12, 1964, on S. 2573. 88th Cong., 2nd sess.

Vaidyaraman, P. P., and Brahme, S. B., 1977. "Tidal power generation in India," *Inst. Eng. J.(India)* part EL 57:200–206.

Valembois, J., 1957. Possibilité de captage de l'énergie de la houle au moyen de résonateurs: *La Houille Blanche* (IV$^{es}$ Journ. de l'Hydraulique) II 418–420.

Valembois, J., and C. Birard, 1954. Les ouvrages résonants et leur application à la protection des ports: *Proc. 5th Conf. Coast. Eng.(Berkeley)*.

Vallarino, E., and Castillo, C., 1960, Evaluacion del potencial mareomotreiz de las costas españolas. *World Power Conf.* (Madrid) II, C-16.

Van London, A. M., 1954. *The mode of action of anti-fouling paints: interaction between anti-fouling paints and sea water*. Amsterdam, Netherlands: Paint Research Institute (Report 62C).

Vantroys, L., 1957a. Nature de l'énergie des marées: *La Houille Blanche* (IV$^{es}$ Journ. de l'Hydraulique) I, 133–141.

Vantroys, L., 1957b. "Le régime des maérés dans la Manche," *La Houille Blanche* (IV$^{es}$ Journ. de l'Hydraulique) I, 176–181.

Vantroys, L., 1957c. Perturbation apportée au régime des marées par le fonctionnement d'une usine marémotrice, *La Houille Blanche* (IV$^{es}$ Journ. de l'Hydraulique) **I**, 188–199.
Vantroys, L., 1958. Le remous d'un ouvrage dans une mer à marée: *Bull. Inf. Comité Central d'Océanogr.* **X**, 8, 9, and 10.
Vernon, K. R., 1974. Hydro (including tidal) energy: *Philos. Trans. Roy. Soc. London* Ser. A, 276, 485–493.
Vincent, M., 1924. *Réflections sur l'utilisation future des énergies naturelles: vagues, chutes hydrauliques et barométriques, chaleur solaire.* Paris: Fischbacher.
Voss, A., 1979, Waves, currents, tide-problems and prospects: *Energy* 4, 5, 823–831.
Voyer, M., and Penel, M., 1957. Les calculs de la production d'une usine marémotrice, *La Houille Blanche* (IV$^{es}$ Journées de l'Hydrauliques) **II**, 472–487.
Wailes, R., 1941. Tide mills in England and Wales: *Junior Inst. Eng., J. and Record of Trans.*, **51**, 91–114.
Walker, H., 1965. France meets the sea in Brittany, *National Geographic J.* **CIII**, 4 (April).
Waller, D. H., 1970. Environmental effects of tidal power development. In Gray and Gashus (eds.), *Tidal Power.* New York: Plenum, pp. 611–625.
Warnock, J. C., and Tanner, R. G., 1978. Selection of optimum sites for tidal power development in the Bay of Fundy, *Proc. 1st Int. Symp. Wave & Tidal Energy (Canterbury)* **I**, 1, E1, 1–22.
Witherell, C. and Debelius, C. A., 1981, Preliminary assessment of Cook Inlet tidal power: *Proc. 2nd Int. Symp. Wave and Tidal En.* (Cambridge) II, 421–435.
Wayne, W. W., Jr., 1977a. *Tidal power study for the U.S. Energy Research and Development Administration,* final report. Boston: Stone and Webster Engineering Corp., 2 vols.
Wayne, W. W., Jr., 1977b. The current status of tidal power: *Key Biscayne (Florida), Nat. Symp. on Energy and the Oceans,* Conf. Coursebook 7–58.
Wayne, W. W., Jr., 1978. Tidal power possibilities in the United States, *A.S.C.E.*, Pittsburgh Preprint 3199.
Wayne, W. W., 1981, North American tidal power prospects: *The International Journal of Ambient Energy* **2**, 3, 10–19.
Wertheim, J. K., 1961. Studies on the electrical potential between Key West, Florida and Havana, Cuba, *Trans. Amer. Geophys. Union* 35, 872–875.
Wheathley, J. D., 1977. Impact on social and recreational habits. In Shaw, T. L. (ed.), *An environmental appraisal of the Severn Barrage.* Bristol, The University, pp. 151–154.
Wheeler, S. J., 1981. Optimization of tidal power schemes, *2nd Int. Symp. Wave and Tidal. En.* (Cambridge), 237–248.
Wick, G. L., 1977. Prospects for renewable energy from the sea, *Mar. Techn. Soc. J.* 11, 5–6, 16–21.
Wickert, G., 1976, Tidal power: *Water Power* 8, 6, 221–225 [and] 8, 7, 259–263.
Widdern, H., Cardinal von, 1952/53. La turbine tubulaire, *Escher Wyss Bulletin* **XXV/XXVI**, 22–30.
Wilson, E. M., 1964. A new approach to power from the tides, *New Scientist*, **24**, 415, 290–291.
Wilson, E. M., 1965a. Energy from the tides: *Science Journal* **I**, 5, 50–56.
Wilson, E. M., 1965b. The Solway Firth tidal-power project, *Water Power*, 17, 11, 431–440.
Wilson, E. M., 1965c, A feasibility study of tidal power from Loughs Strangford and Carlingford with pumped storage at Rostrevor. *Proc. Inst. Civ. Eng.* 32, 9, 1–29.
Wilson, E. M., 1966. Feasibility study of tidal power from Loughs Stangford and Carlingford, with pumped storage at Rostrevor, *Inst. Civ. Eng.*, **34**, 83–100.
Wilson, E. M., 1968. The Bristol Channel barrage project: *Proc. Conf. Coastal Eng.* XI, 83, 1304–1325.
Wilson, E. M., 1972. Tidal energy and its development. In *Conf. on Engineering in the Ocean Environment,* 3d ed. *Ocean '72.* New York: Institute of Electrical and Electronic Engineers, 30, 47–56.
Wilson, E. M., 1973. Energy from the sea—tidal power, *Underwater J.* 5, 4, 175–186.
Wilson, E. M., 1977. Tidal energy and system planning, *Consulting Engineer (London)* 41, 4, 25.

Wilson, E. M. and Gibson, R. A. 1978, Studies in retiming tidal energy: *Proc. 1st Int. Symp. Wave and Tidal Energy*, I, 1, H1, 1–10.
Wilson, E. M., Severn, B., Swales, M. C., Henery, D., 1968. The Bristol Channel barrage project. *Proc. Int. Conf. Coast. Engng.* (London) XI, 1304–1325.
Wilson, E. M., and Swales, M. C., 1972. Tidal power from Cook Inlet, Alaska. In Gray and Gashus (eds.), *Tidal Power*. New York: Plenum, pp. 239–248.
Winters, A., 1972. A desk study of the Severn Estuary, *Proc. Math. & Hydraul. Model. of Estuar. Poll., Water Poll*. Technical Paper No. 13, 105–113.
Wishart, S. J., 1981. A preliminary survey of tidal energy from five U.K. estuaries, *2nd Int. Symp. Wave and Tidal En.* (Cambridge), 299–314.
Won, T. S., 1975. The tidal power resources and their power generation projects of the western coast of Korea, *Proc. Pacif. Sci. Assn., Pac. Sci. Cong.* **XIII**, I, 162.
Won, T. S., 1977. Tidal power projects on the west coast of the Republic of Korea. *World Energy Conf.* (Istanbul) Sept.
Yen, J. T., and Isaacs, J., 1978. Dynamic dam for harnessing ocean and river currents and tidal power, *Proc. Ann. Mar. Techn. Soc./IEEE Comp. Conf., Ocean '78: The ocean challenge.* **IV**, 582–584.
Yong, W. J., 1977. Tidal power projects on the west coast of the Republic of Korea, *World Energy Conf.(Istambul)* **X**, 1–11.

## *Additional Anonymous works listed by year of publication*

1899. Les moulins à marée de New York, *Revue Scientifique*, **IV**, 11, 1, 30.
1902. Le moteur à marée en Californie, *Revue Scientifique*, **IV**, 17, 8, 253.
1929. *Utilizacion de las mareas de las costas Patagonicas*: Buenos Aires.
1933. *Report of the Severn Barrage Committee*. London: H. M. Stationery Office.
1933. *Appendix to the Brabazon Severn Barrage committee report: Report of the expert coordinating subcommittee*. London: H. M. Stationery Office.
1934. A tidal power project at Avonmouth, *Energelicheskoc Obozvenie* 7.
1941. *Passamaquoddy Tidal Power Project, Maine*. Washington, D.C.: Federal Power Commission.
1945. *Report on the Severn barrage scheme*, London, H.M. Stat. Off.
1945. *Report on tidal power, Petitcodiac and Memramcook estuaries*. Frederictown: Government of the Province of New Brunswick.
1946. *Rapport sur les études par la Société d'Etudes pour l'Utilisation des Marées*. Paris: Electricité de France.
1946. *Report on tidal power, Petitcodiak, and Memramcook estuaries, Province of New Brunswick*. Ottawa: King's Printer.
1958. Tidal power. Argentina signs contract for study: *Engng. News Record* 160, 1, (Oct. 19), 56.
1959. The Dutch Delta Plan: *The Engineer* 208, 5410, 337–352.
1959. Passamaquoddy feasible for United States, engineers say, *Electrical World*, **152**, 54.
1959. *Proc. of the All-Chinese Conf. on Tidal Power Utilization*: Shangai, Pamphlets.
1962. The Rance tidal power plant, *La Houille Blanche* **XV**, 2, 117–129.
1963. The closure of the Rance estuary, *La Houille Blanche* **XVIII**, 7, 789–798.
1964. The estuarine barrage, *Engineer*, **218**, 5677, 787–788.
1964. "The Quoddy question: time and tide," *IEEE* I, 9, 96–118.
1964. Constructors harness the tides in France, *Engineering News Record* **173**, 11, 32–34.
1965. White House offers Marine Powers Plan, *New York Times* July 11, 1 and 51.
1965. Power and water, *Congressional Record* (August 17), A4580 (daily edition).
1965. Editorial from Patriot Ledger, Quincy, Mass.
1965. Border power project announced for Maine, *Washington Post* July 11, A6.
1966. Tidal power from the Bristol Channel, *Engineer* **222**, 5739, 109–110.

1966. Tidal power comes to France, *Engineer* **202**, 5228, 17–24.
1966. The Rance tidal-power station, *Engineer* **222**, 5784, 856–860.
1966. The Rance tidal-power station, *Engineer* **222**, 5785, 891–895.
1966. France harnesses sea tides for electrical power, *France Actuelle*, **XV**, 16.
1967. Canada sparks Fundy tidal power study, *Engineering News Record* 179, 22, 25–27.
1967. Power from the Rance and the Rhine, *Engineer* **223**, 5790, 74–75.
1967. Tidal power from the Severn, *Engineer* **223**, 5802, 509 and 513.
1967. Bay of Fundy tidal power study, *Engineer* **223**, 5809, 786.
1967. Rance tidal power plant, *Materials Protection* **VI**, 1, 46–47.
1967. France harnesses the tides, *Ocean Industry* **II**, 2, 5–8.
1967. The Rance tidal power scheme, *Water Power* **19**, 1.
1967. Rance tidal power plant, *Materials Protection* **6**, 1, 46–47.
1968. English electric and tidal power, *Engineer* **226**, 5877, 382.
1969. *Handbook of ocean and underwater engineering*. New York: McGraw-Hill.
1969. *Feasibility of tidal power development in the Bay of Fundy. Board and Committee Report*. Ottawa: Atlantic Tidal Power Programming Board.
1969. *Feasibility of tidal power development in the Bay of Fundy*. Ottawa, Atl. T.O. Programm. Bd.
1973. Six ans d'exploitation de l'usine marémotrice de la Rance, *La Houille Blanche* **211**, 2–3, 1–66; 125–270.
1973. *An energy policy for Canada. Summary of analysis*. Ottawa: Dept. of Energy, Mines & Resources.
1974. Tidal power may now make sense, *Business Week* **2356**, 115–118.
1975. The Severn Barrage: its possible implications, *Estuar. & Brackish Water Sci. Assn.* **11**, 7, 8–14.
1976. *Severn Barrage study*. Hydraulics Research Station.
1976/7. *Water resources development project, St. John River Basin, Dickey-Lincoln School Lakes design. Memo. Nos. 3–4A*. Walton, Mass.: Army Corps of Engineers, New England Division.
1977. Severn barrage seminar. *Proc. Energy* Paper 27, London H.M. Stat. Off. 1977. *Reassessment of Fundy tidal power*. Bay of Fundy Tidal Power Review Board, Ottawa.
1977. *The energy source book:* Aspen, Colorado, The Center for Compliance Information, Aspen Systems Corp. (Part 3).
1977. *Tidal power barrages in the Severn estuary. Recent evidence of their feasibility*. London: H.M.S.O. (Energy Paper No. 23).
1976. L'usine marémotrice de Chausey. In *La production d'électricité d'origine hydraulique. Rapport de la Commission de la production d'origine hydraulique et marine*. Paris: La Documentation Française **IV**, 41–49.
1977. Reassessment of Fundy tidal power. Ottawa, Bay of Fundy Tid. Pow. Rev. Bd.
1977. British scheme to harness the tides may yet go ahead, *Ocean Industry* **12**, 8, 107–109.
1978. *Severn Barrage seminar*. September 7, 1977. London: H.M.S.O.
1980. Electric power for China's modernisation: the hydroelectric option. Washington D.C., C.I.A.
1981. Canadians utilize hydroelectric power. *J. of Commerce* (Feb. 25).

# 1982–1992

Aisicks, E. G. and Zyngierman, I., 1984, The San José Gulf tidal power plant, Argentina: *Proc. ECOR '84 and 1st Alt. En. Argent. Conf.* **II**, 1–9.
Anonymous, 1982, Tidal power generation (in Japanese): *Journal of the Society of Naval Architects of Japan* July, no page numbers.
Anonymous, 1983. *Mersey barrage pre-feasibility study research report*: London, Marintech North West.
Anonymous, 1985. *Mersey barrage, exploratory barrage tests*: London, Hydraulics Research Ltd.

Anonymous, 1987a, 6 bn. tidal barrage economic: *Int. Constr.* X, 1987, 15.
Anonymous, 1987b, Catching the tide power: *Int. Constr.* X, 1987, 7.
Anonymous, 1987c, *Tidal power*: London, Telford.
Antonioli, A., 1982, Tidal Power: *Int. J. Hydrogr. Energy* VII (6), 507–517.
Baird, H., 1982, Upgrading the low head of low density hydraulic energy, making it more acceptable for putting through a conventional hydraulic turbine: Proc. Int. Conf. New Approach: *New Approaches to Tidal Power Conf.*, Bedford Inst. Oceano. Proc., 10, 1–15.
Baker, G. C., 1982a, Tidal Power. Some historical implications: *New Approaches to Tidal Power,* Conf. Bedford Inst. Oceanog. Proc., 1, 1–6.
Baker, A. C., 1982b, Overview of environmental aspects of the Severn development: *Proc. Int. Conf. New Approaches to Tidal Power*, Bedford Inst. Oceanogr. Dartmouth (N.S.) 9, 1–11.
Baker, A.C., 1987, Tidal power: *IEEE Proc.* [section A] 134, 5, 392–398.
Baker, A. C., 1991, *Tidal power*: London, Peregrinns Baker, C., 1991.
Tidal power: *Journal of Energy Policy* 19, 8, 792–797.
Baker, A.C. and Wishart, S.J., 1986, Tidal power for small estuaries: *Proc. 3rd Int. Symp. on Wave, Tidal, OTEC and small scale hydro energy* 115–123.
Banal, M. and Bichon, A., 1982, Tidal energy in France – The Rance tidal power station. Some results after 15 years of operation: *J. Inst. Energy* 55, 423, 86–91.
Banal, M., 1982, Tidal energy in 1982: *La Houille Blanche – Rev. Int. de l'Eau* 37, 5 & 6, 433–439.
Bellamy, N. W., 1986. A pneumatic low-head hydroelectric system, *Proc. 3rd Int. Symp. Wave et al.* III, 31–42.
Bickley, D. T. and Ryrie, S. C., 1982, A two-basin tidal power scheme for the Severn Estuary: *Int. Conf. New Approaches to Tidal Power* (Dartmouth, N.S.), 12, 1–32.
Birket, N., Count, B.M. and Nichols, N. K., 1984, Optimal control problems in tidal power. *Int. Water Power and Dam Constr.* 36, 1, 37–42.
Bonnefille, R., 1982, Tidal power stations. In Thielheim, K.O. (ed.), *Primary energy. Present status and future perspectives:* Heidelberg, Springer Verlag pp. 319–334.
Booda, L.L., 1985, River Rance tidal power-plant nears 20 years in operation: *Sea Technology* 26, 9, 22.
Bosc. J., Megrint, L., 1984. Evolution des groupes axiaux pour l'équipment des installations marémotrices. *La Houille Blanche* 8, 590–595.
Breeze, P., 1987, When will fortune turn for tidal power?: *Modern Power Syst.* 7, 3, 19–25
Bullock, G. N., Wilson, E. M., 1982, The Salford transverse oscillator: *New Approaches Tidal Power Conf.*, Bedford Inst. Oceanog. Proc., 13, 1–8.
Carr, G. R., 1986. Studies of a tidal power barrage on the Rivery Mersey. *Proc. Water for energy, 3rd Int. Symp. Wave et al.* Paper 31, 1–23.
Cave, P. R. and Evans, E.M., 1984. Tidal energy systems for isolated communities. In West, E. (ed.) *Alternative energy systems*: New York, Pergamon XI +351 pp. 9–14.
Cave, P. R. and Evans, E.M., 1985, Tidal streams and energy supply in the Channel Islands. *Proc. Conf. for Rural and Island Communities* (Inverness).
Cave, P. R. and Evans, E.M., 1986, Tidal stream energy economics, technology and cost: *3rd Int. Symp. Wave, Tidal, OTEC and small scale hydro energy "Water for Energy"* 133–186.
Charlier, R. H., 1982a, Oceans and electricity: *Int. J. Environmental Studies* 18, 3/4, 159–168 (Pt.1); 19, 1, 7–16 (Pt. II).
Charlier, R. H., 1982b, *Tidal energy*: New York, Van Nostrand-Reinhold.
Charlier, R. H., 1982c, Developments in tidal power: *Intersoc. Energy Conversion Engng. Conf. Proc.* XVII
Charlier, R. H., 1983, Small ocean-powered schemes, in: Meyer and Olson, eds., *The future of small energy resources*, New York, Mc Graw-Hill.
Charlier, R. H., 1984, *Geostatistics. Class Notes*: Chicago, Northeastern Illinois University 104 pp.
Charlier, R. H., 1990, Ocean energy. Historical Development of its harnessing: *Proc. 4th Int. Congr. Hist. Oceanog.* Sondermuinmer, Deutsche hydrogen, Zeitung 22, 499–503.
Clavier, J., 1981, *Thèse*: Université de Paris-IV.

Cotillon, J., 1974, La Rance: six years of operating a tidal power plant in France: *Water Power* 26, 10, 314–322.
Cotillon, J., 1977a, Advantages of bulb-units for low-head developments: *Int. Water Power & Dam Constr.* 29, 1, -26
Cotillon, J., 1977b, The advantages of the tube-turbine sets for low-head hydro-electric plants: *Elektroprivreda* 30, 5–6, 189–195 [in Serbian]
Couzet, P. and Boissart, P., 1978, Pollution of the Rance estuary by effluent from St Malo and Dinard: *La Houille Blanche – Rev. Int. de l'Eau* 33, 7–8, 561–567.
Daborn, G. R., 1982, Biological consequences of the Minas Basin Tidal Power proposal: *Proc. Int. Conf. New Approaches to Tidal Power*, Bedford Inst. Oceanogr. Dartmouth (N.S.), 11, 1–14.
Dauvin, J. C., 1984, *Thèse*: Paris, Université Pierre et Marie Curie.
Davis, B. V., Swan, D. H., Jeffers, K. A., Farrell, J. R. and Calder, L. M., 1982, *Ultra-low head hydroelectric power generation using ducted vertical axis water turbines*: Ottawa, Nat. Res. Hydr. Lab.—Nova Energy Ltd. Report NEL–022.
Davis, C. V. and Swan, A. H., 1982, Extracting energy from river and tidal currents using open and ducted vertical axis turbines. Model tests and prototype design: *New Approach. Tidal Power, Conf.* Bedford Inst. Oceanogr. Proc.
de la Vallée Poussin, S., 1991, Automatic control of the La Rance tidal power plant : *Rev. Gén. de l'Electricité* [1991] 3, 31–40.
DeLory, R. P., 1987, Prototype tidal power plant achieves 99% availability: *Sulzer Techn. Rev.*1, 3–8.
DeLory, R. P., 1986. The Annapolis tidal generating station. *3rd Int. Symp. Wave et al.*, 125–132.
Design Institute of the Yao-Ning Province Administration of Water Management and Power: Tidal Power.
Fay, J. A., 1982, Design principles of horizontal-axis tidal current turbines: *New Approach. Tidal Power Conf.*, Bedford Inst. Oceanogr. Proc.
Fay, J. A. and Smachlo, M. A., 1982, *Small scale tidal power plants*: Cambridge, MA., Massachusetts Institute of Technology (MIT Sea Grant College Program).
Gilbert, R. L. G., 1982a, Retiming with hydrogen: *New Approach. Tidal Power Conf.*, Bedford Inst. Oceanogr. Proc., 5, 1–4.
Gilbert, R., 1982b, Designing for inefficiency: *Proc. Int. Conf. New Approaches to Tidal Power*, 5, 5–8.
Gorlov, A. M., 1982, Hydropneumatic Approach to Harnessing Tidal Power: *New Approach. Tidal power Conf.* Bedford Inst. Oceanogr. Proc.
Greenberg, D., 1982, Environmental effects: *New Approach. Tidal Power Conf.* Bedford Inst. Oceano. Proc.
Guixiang, L., 1991, Prospects for the resources on the tidal energy development in China: *Collection of Oceanographic Works-Haiyang Wenji* 14, 1, 128–134.
Haydock, J. L. and Warnock, J. G., 1982, Energy storage for tidal plants using compressed air: *New Approach. Tidal Power Conf.*, Bedford Inst. Oceanogr. Proc.
Han, S. J., 1982, *Thèse:* Université de Paris-Sud.
Han, S. J., 1983, Sedimentary conditions in tidal power plant areas of Garolim Bay, Korea and Rance, France: *Bull. Kor. Res. Dev. Inst.* 5, 1, 27–34.
Hilton, D. J., 1982, Performance of vertical axis water turbine as a tidal energy converter: *Proc. Int. Conf. New Approaches to Tidal Power*, Bedford Inst. Oceanogr., 7, 1–10.
Hollenstein, M. and Soland, W., 1982, The bulb turbines for the Rance power station: *Escher Wyss News*, 54/55.
Hillariet, P., 1989, The Rance tidal power plant: *Rev. de l'Energie* 40, 410, 264–270.
Hillariet, P. and Weisrock, G., 1986, Optimizing production from the Rance tidal power station: *3rd Int. Symp. Wave, Tidal, etc: Water for Energy* 165–177.
Kato, N. and Ohashi, Y., 1984, A study of energy extraction system from ocean and tidal currents: *Proc. Int. Conf. ECOR '84 and 1st Argent. Ocean Eng. Congr.* (Buenos Aires Oct.1984) I, 115–132.

Keefer, B. G., 1982, Optimized low head approach to tidal energy: *New Approach. Tidal Power Conf.*, Bedford Inst. Oceanog. Proc.

Kinno, H., Zielke, W. et al., eds., 1982, A microcomputer program for design of the self-retiming mechanics of large tidal power sites: *4th Int. Conf. Finite Elem. in Water Res.* (Hannover, BRD).

Kristoferson, L. A. and Bokalders, V., 1991, *Renewable energies technologies. Their applications in developing countries*: Rugby, Intermediate Technology Publications, 336 pp.

Lacalvez, J. C., 1986, *Thèse:* Université de Rennes-1.

Lang, F., 1986, *Thèse:* Université de Rennes-1.

Lechapt, P.; 1986, *Thèse:* Université de Rennes-1.

Megnint, M.L. and Allegre, M.J., 1986b, Present design of tidal bulb-units based on the experience in the Rance tidal power plant: *3rd Int. Symp. Wave, Tidal, OTEC and Small Hydro-Energy*: Water for Energy: 147–164.

Ménanteau, L., 1991. Zones humides du littoral de la Communauté européenne vues de l'Espace/Wetlands of the European Community seen from space/Zonas humedas de la comumidad europea vistas desde el espacio. Conférence des Régions Périphériques Maritimes de la C.E.E. [Trilingual] 184 pp.

Ménanteau, L., Guillemot, E. and Vamey, J.-R., 1989. Mapa fisigrafico del litorol atlántico de Andalucía/Carte physiographique du littoral atlantique del'andalousie/Physiographic map of the atlantic littoral of Andalucía. M.F. 04 (Rita La Barrosa – Baliza de cadiz).

Merlin, A., Sandrin, P., Gres, J.M. and Hillariet, M., 1982, AGRA, the new operation model for the La Rance tidal power-plant:*IEEE Trans. Power Apparatus & Syst.*101, 2, 290–294.

Mettam, C. J., 1982, Tidal power from the Severn estuary: an essay review: *Nat. Wales* 1, 1, 50–61.

Montalverne, G 1991. Moinhos de maré. Atlantis 1, 91, 67–71.

Moore, G., 1985, Time for tide? (tidal power): *Electron and Power*, 32, 11, 823–826.

Mourier, M., Digue, C., Lecouturier, J. and Piat, H., 1990, Automatic control installed at La Rance: *Int. Power & Dam Constr.* 42, 1, 28 & 34–36.

Orava, P.J. and Eirola, T.J., 1983, On the optimization of the operation of tidal power plants: application of the back-and-forth shooting method: *Acta Polytechnica Scandinavia* [Applied Physics Series] Ph138, 152–158.

Peatfield, A. M. et al., 1984. Energy from low-head water sources. In: West, E. (ed.), *Alternative energy systems*, London, Pergamon pp. 1–8.

Pratte, B. D., 1982a, Theory and comparison of vertical and horizontal axis turbines including ducting: *New Approach. Tidal Power Conf.*, Bedford Inst. Oceanogr. Proc., 1982.

Pratte, D. B., 1982b, Overview of environmental aspects of the Severn development: *Proc. Int. Conf. New Approaches to Tidal Power*, Bedford Inst. Oceanogr. Dartmouth (N.S.) 9, 1–11.

Pratte, D. B., 1982c, Overview of non-conventional current energy conversion systems: *Proc. Int. Conf. New Approaches to Tidal Power*, Bedford Inst. Oceanogr. Dartmouth (N.S.), 4, 1–8.

Rivain, V., 1983, *Thèse:* Université de Paris-VI.

Robinson, I. S., 1982, Enhancement of power availability by tuning the operation of the tidal barriers to match the tidal dynamics: *Int. Conf. New Appr. to Tidal Power*, Dartmouth NS, Bedford Inst. of Ocean. 1–2.

Rolfe, R. 1987, Taming the tide: *Fin. Times*, Weekend, 4.4.87, I.

Russell, E., 2004, Could the tide be turning? (Tidal power generation): *Int. Power Gener.* 27, 3, 24–27.

Scarratt D. J., 1982, Possible effects of various tidal power schemes on local and migrant fish populations in the Bay of Fundy: *Proc.New Approach. Tidal Power Conf.*, Bedford Inst. Oceanogr.

Searratt, D. J., Dadswell, M. J., 1982. The effects of tidal power generating systems on local and migrant fish populations. *Proc. Conf. New Appr. to Tidal Power* (Dartmouth NS) 6, 1–20.

Severn Barrage Committee, 1983, *Tidal power from the Severn estuary*: London, HMSO.

Severn Tidal Power Group, 1986, *Tidal power from the Severn*: London, Telford.

Sharma, H. R., 1982, India embarks on tidal power: *Water Power and Dam Construction* 34, 6, 32.

Sharma, H. R., 1986. Kachchh tidal power project. *3rd Int. Symp. Wave, Tidal et al*: III, pt. 14, 179–189.

Sharma, H. R., Vats, T. P., 1985. Environmental impacts of the proposed Kachchh tidal power plant: *Proc. Int. Sem. Env. Impact Assessm. Water Res.* (Roorkee, India).
Smachlo, M. A., 1982, *A generic model of small-scale tidal power plant operation and performance* (Master's Degree Thesis): Cambridge, MA., Massachusetts Institute of Technology.
Song, W. O. and Van Walsum, E., 1979, Korea tidal power and beyond: *Proceed-ings, 2nd Miami Int. Conf. on Alternative Energy Sources* (Miami).
Song, W. O., 1983, KORDI's contribution to Korean tidal power studies: *Bull. Kor. Res. Dev. Inst.* 5, 1, 47–54.
Song, W. O., 1987, Reassessment of Garolim tidal power project: *Oc.Res.* (Korea) 9, 1/2, 29–33.
Stam, B., 1987, Getijdencentrale is voor Nederland een reeds lang gepasseerd station: *PT/Aktueel* 14, 1.4.87, 15.
Stanbury, J., 1986, Optimizing production from the Rance tidal power station: *3rd Int. Symp. Wave, Tidal, OTEC and Small scale Hydro-energy: Water for Energy* 165–177.
Stamberger, A., 1991, Large tidal generating stations: *Electrotechnik* 44, 5, 39–41.
Twidell, J. and Weir, T., 1990, *Renewable energy resources*: London, Spon.
Vaidyraman, P. P. et al., 1986. Experience during data collection for tidal power project in the Gulf of Kachchh: *3rd Int. Symp. Wave, Tidal et al.* III, paper # 15, 191–201.
Vanney, J. R., and Ménanteau, L., 1985. Mapa fisiográfico del litoral atlántico de Andalucía/Carte physiographique du littoral atlantique de l'andalousie/Physiographic Map of the atlantic littoral of andalucia. M.F.02. Punta Umbría – Matalascañas; M.F.03. Matalascañas – Chipiona. Casa de velácquer and Junta de andalucia.
van Walsum, E., 1982, Tidal power in the Bay of Fundy: *Proc.New Approach. Tidal Power Conf.*, Bedford Inst. Oceanog.
White, P. R. S. et al., 1985. A low head hydro scheme suitable for small tidal and river applications. *Proc. Int. Conf. Energy for Rural and Island Comm.* (Inverness).
Wilson, E.M., 1983, Tidal power reviewed: *Int. Water Power & Dam Constr.* 35, 9, 13–16.
Zhuang, J., 1991, Exploring study of optimum patterns of the tidal power plant: *Ocean Engineering-Haiyung Gongeheng* 9, 2, 82–90.
Zu-Tian, G., 1989, The development of tidal resources in China, in Krock, H. J., *Ocean Recovery, ICOER 89, Proceedings Int. Conf. Ocean Energy Conversion: ASCE, Water, Ports & Coasts, Ocean Div.*, New York, The Society, pp. 157–166.

# What is Being Said: 1992–2007

Alnaser, W.E., 1995, Renewable energy resources in the State of Bahrain: *Applied Energy* **50**, 1, 23–30.
Anonymous, 1997, The Rance River tidal power station – Discussion: *La Houille Blanche – Rev. Int. de l'Eau* 52, 3, 108.
Anonymous, 1997a, The Rance River tidal power station184–191. – Discussion: *La Houille Blanche – Rev. Int. de l'Eau* 52, 3, 109–110.
Anonymous, 1997b, The Rance tidal power plant—Discussion: *Houille Blanche– Revue Int. de l'Eau* 52, 3, 108.
Anonymous, 1997c, The Rance tidal power plant)—Discussion: *Houille Blanche- Rev. Int. de l'Eau* 52, 3, 109–110.
Anonymous, 1998, Tidal power gets new backing: *Chemical Engineering* (London): 4, 655, 12–13.
Anonymous, 2002a, Renewables – Tidal energy: *Chemistry & Industry* [2002], 2, 10.
Anonymous, 2002b, Utility buys into tidal power: *Profess. Engng* 15, 3, 7.
Anonymous, 2003a, Gas know-how helps harness tidal power: *Int. Gas Engng and Man.* 43, 4, 43.
Anonymous, 2003b, Norway nabs tidal-power crown: *Power* 147, 2, 6.

Anonymous, 2003c, Ocean energy development: obstacles to commercialization: *Oceans 2003* 4, 2278–2283.
Anonymous, 2003d, Tidal energy gaining power in Norway: *Bull. Am. Meteorol. Soc.* 84, 1, 11–12.
Anonymous, 2004, Plugging into the ocean (Electric power generation): *Machine Design Int.* 76, 18, 82–90.
Anonymous, 2005a, Cash grant should put prototype tidal-power generator in the swim: *Profess. Engng* 18, 11, 10.
Anonymous, 2005b, *Proc. Gen. Meet. IEEE Power Engng Soc.* 3 vols 3093 pp.
Anonymous, 2005c, Tidal power making waves: *TCE-Inst. Chem. Eng*.765, 9.
Anonymous, 2006, Tidal power trials in New York's East River: *TCE-Inst. Chem. Eng.* 760, 9.31, 2, 173–180.
Appleyard, D., 2005, High tide (tidal power): *Int. Power & Dam Constr.* 57, 4, 36–37.
BabusHaq, R. F. and Probert, S. D., 1996, Combined heat-and-power implementation in the UK: Past, present and prospective developments: *Applied Energy* 53, 1–2, 47–76.
Bahaj, A.S. and Myers, L.E., 2004a, Analytical estimates of the energy yield potential from the Alderney Race (Channel Islands) using marine current energy converters: *Ren. Energy* 29, 12, 1939–1945.
Bahj, A.S. and Myers, L.E., 2004b, Fundamentals applicable to the utilization of marine current turbines for energy production: *Renewable Energy* 28, 14, 2205–2211.
Banal, M., 1997a, History of tidal power in France: *La Houille Blanche-Journal International de l'Eau* 52, 3, 14–15.
Banal, M., 1997b, The technical origins of the tidal power station: *La Houille Blanche-Rev. Int. de l'Eau* 52, 3, 16–17.
Barreau, M., 1997, 30th anniversary of the Rance tidal power station: *La Houille Blanche-Revue Internationale de l'Eau* 52, 3, 13.
Bartle, A. and Hallowes, G., 2005, Hydroelectric power: present role and future: *Proc. Instit. Civ. Eng.-Civ. Engng* 158, sp.iss. 2, 28–31.
Bedford Inst. Oceangr. Dartmouth (N.S.), 1982, *Proc. New Appr. Tidal Power.*
Belayev, L.S., Chudinova, L.Y., Koshcheev, L.A., Podkovalnikov, S.V, Savelyev, V.A. and Voropai, N.I., 2003, The high voltage direct current bus "Siberia-Russian Far East": *Proc. Gen. Meet. IEEEPower Engng Soc.* 4, 2186–2190.
Bernshtein, L. B. and Usachev, I. N., 1997, Utilizaton of tidal power, and Russia in overcoming the global and ecological crisis: *La Houille Blanche-Revue Int. de l'Eau* 52, 3, 96–102.
Bernshtein, L. B., 1995, Tidal power development. A realistic, justifiable and top-ical problem of today: *IEEE Transactions on Energy Conversion* 10, 591–599.
Bhaj, A.S. and Myers, L., 2003, Fundamentals applicable to the utilization of marine current turbines for energy production: *Renewable Energy* 28, 14, 2205–2211.
Bitterlin, I.F., 2005, A review of the future for wind and tidal power generation in the UK: *Proc. 5th IASTED Int. Conf. Power & Energy Syst.* 184–191.
Blunden, B.S. and Bahaj, A.S., 2006, Initial evaluation of tidal energy resources at Portland Bill, UK: *Renewable Energy* 31, 2, 121–132.
Brooks, D. A., 1992, Tides and tidal power in Passamaquoddy Bay—A numerical simulation: *Continental Shelf Research* 12, 5–6, 675–716.
Bryans, A.G., Fox, B., Crossley, P.A. and O'Malley, M., 2005, Impact of tidal generation on power system operation in Ireland: *IEEE Trans. Power Syst.* 52, 9 & 10, 1075–1086; 20, 4, 2034–2040.
Bryans, A.G., Fox, B., Crossley, P.A. and Whittaker, T.J.T., 2004, Tidal energy resource assessment for the Irish grid: *Proc. 39th Int. Universities Power Engng Conf.* 1, 1, 614–617.
Bryden, I., 2002. The future of tidal current power: "Oceans of change" *Int. Conf.*
Bryden, I.G., 2003, Tidal power on the horizon: *Profess. Engng* 16, 15, 22–23.
Bryden, I.G. and MacFarlane, D.M., 2000, The utilisation of short term energy storage with tidal current generation systems: *Energy* 25, 9, 893–907.
Bryden, I. G. and Melville, G.T., 2004, Choosing and evaluating sites for tidal power development: *Proc. Inst. Mech. Eng.* [Pt A, J. Power and Energy]: 219, A–3, 235–247.

# Annex I: General Bibliography

Bryden, I.G., Grinsted, T. and Melville, G.T., 2004, Assessing the potential of a single tidal channel to deliver useful energy:*Appl. Oc. Res.* 26, 5, 198–204.
Chaineux, M.C. and Charlier, R.H., 2008, Women and tidal power. *Ren. Sust. En. Rev.* (in press)
Chang, J., Leung, D.Y.C., Wu, C.Z. and Yuan, Z.H., 2003, A review of the energy production, consumption, and prospect of renewable energy in China: *Ren. Sust. Energy Rev.* 7, 5, 453–468.
Charlier, R. H. and Justus, J. R., 1993, *Ocean energies. Environmental, economic and technological aspects of alternative power sources*: Amsterdam-London-New York-Tokyo, Elsevier, 665 pp.
Charlier, R. H. and Ménanteau, L., 1998, The saga of tide mills: *Renewable and Sustainable Energy Reviews* 1, 3, 1–44.
Charlier, R. H., 1998, Re-invention or aggiornamento? Tidal power at 30 years: *Renewable and Sustainable Energy Reviews* 1, 4, 271–289.
Charlier, R. H., 2001, Ocean alternative energy. The view from China—"small is beautiful": *Renewable and Sustainable Energy Reviews* 5, 4, 403–409.
Charlier, R. H., 2003a. Energy from waves, winds, and tides: *Proc. XXXth Convocation Pacem in Maribus*, 2003.
Charlier, R. H., 2003b. Tidal power: in Schwartz, M. (ed.), *Enc. Mar. Sci.* Heidelberg, Springer, pp. 1437–1442.
Charlier, R.H., 2003c, Sustainable co-generation from the tides – A review: *Ren. Sust. Energy Rev.* 7, 3, 187–213.
Charlier, R.H., 2003d, Sustainable co-generation from the tides: Bibliography: *Ren. Sust. Energy Rev.* 7, 3, 215–247.
Charlier, R.H., 2003e, Small is beautiful – The view from China: *Ren. Sust. En. Rev.* 7.5, 445–453.
Charlier, R.H. 2004, a sleeper awakes: power from tidal currents: *Res. Surt. En. Rev.* 8
Charlier, R.H., 2006, *Renew. En.* 2006.
Charlier, R.H., 2007. Forty candles for the Rance River TPP. Tides provide renewable and sustainable power generation. *Ren. Sust. En, Rev.* 11, 2032–2057.
Clark, R. H., 1997, Prospects for Fundy tidal power: *La Houille Blanche-Revue Int. de l'Eau* 52, 3, 79–88.
Clarke, J.A., Connor, G., Grant, A.D. and Johnstone, C.M., 2006, Regulating the output characteristics of tidal power current stations to facilitate better base load matching over the lunar cycle: *Renewable Energy* 31, 2, 173–180.
Dadswell, M. J. and Rulifson, R. A., 1994, Macrotidal estuaries: a region of collision between migratory marine animals and tidal power development: *Biological Journal of the Linnean Society* 51, 1–2, 93–113.
Duckers, L., 1995, Water power. Wave, tidal and low head hydro technologies: *Power Engineering Journal* 9, 4, 164–172.
Fahmy, F.N., 2001, An optimal operation and mathematical model of tidal energy system at Red Sea area: *Proc. 5th Int. Conf. on Electr. Machines & Syst.* [ICEMS-2001] 1, 664–667.
Flin, D., 2005, Sink or swim [tidal power generation]: *Power Eng.* 19, 3, 32–35.
Foreman, M. G. G., Walters, R. A., Henry, R. F., Keller, C. P. and Dolling, A. G., 1995, A tidal model for eastern Juan de Fuca Strait and the southern Strait of Georgia: *Review of Aquatic Science* 100, C1, 721–740.
Fox, P., 2004, Designs on tidal turbines: *Power Eng.* 18, 4, 32–33.
Fraenkel, P.L., 2001, Is the tide turning for marine current turbines? : *Modern Power Syst.* 21, 6, 37–39.
Fraenkel, P.L., 2002, Power from marine currents: *Proc. Inst. Mech. Eng.* [Pt A][ – *J. Power & Energy* 216, A–1, 1–14.
Frau, J. P., 1993, Tidal energy—Promising projects. La Rance, a successful industrial- scale experiment: *IEEE Transactions on Energy Conversion* 8, 552–558.
Fry, C., 2005, All at Sea (Tidal power):*Power Eng.* 19, 5, 24–27.
Garrett, C. and Cummins, F., 2004, Generating power from tidal currents: *Am. Soc. Civ. Eng – J. Waterways, Port, Coastal and Ocean Engng* 130, 3, 114–118.

Garrett, C. and Cummins, F., 2005, The power potential of tidal currents in channels: *Proc. Roy. Soc. London*, Series A, 461, 2060, 2563–2572.

Gooch, D.J., 2000, Materials issues in renewable energy power generation: *Int. Materials Rev.* 45, 1, 1–14.

Gordon, D. C., 1994, Intertidal ecology and potential power impacts, Bay of Fundy: *Biological Journal of the Linnean Society* **51**, 1–2, 17–23.

Gorlov, A.M., 2003, The helical turbine and its applications for tidal and wave power: *Oceans 2003* 4, 1996.

Grad, P., 2004, Changing tide of power generation: *Eng. Australia* 76, 11, 51–52.

Gross, R., 2004, Technologies and innovation for system change in the UK: status, prospects and system requirements of some leading renewable energy options: *Energy Policy* 29, 10, 1757–1771.

Gupta, R., Jiang, F. and Elaveetil, V., 2001, Feasibility studies of a tidal power plant in Singapore: *Proc. 5th Int. Power Engng Conf.* [IPEC |2001] 1, 67–72.

Hammons, T.J., 1993, Tidal power: *Proc. IEEE* 81, 3, 419–433.

Hammons, T.J., 2005, Energy potential of the oceans: tidal, wave, currents, and OTEC: *Proc. 40th Int. Univ. Power Engng Conf.* 1, 2, 1047–1057.

Hammons, T.J. and Tob, H.K., 2001, Energy and the environment: alternatives in UK for $CO_2$ reduction: *Electr. Power Compon. & Syst.* 29, 3, 277–295.

Hastings, S., 2005, Working with Nature: Tidal power: *Int. Pow. Gener.*28, 3, 27–28.

Haws, E. T., 1997, Tidal power—A major prospect for the 21st century: *Proceedings of The Institution of Civil Engineers—Water, Marine and Energy:* **124**, 1, 1–24.

Haws, E. T., 1999, Tidal power—A major prospect for the 21st century: *Proceedings of the Institution of Civil Engineers—Water, Marine and Energy:* **136**, 2, 117–118.

Hecht, J., 1995, High time for compact tidal power: *New Scientist* **145**, 1964, 20.

Hsu Ku, Cho-Teng, L., Char-Shine, L. and Ming-Kuang, H., 2000, Power generation from Kuroshio east of Taiwan: *Monthly J. of Taipower Engng* 624, 81–89.

ibidem, 2002.

idem: *Proc. 3rd Int. Conf. Energy Coop. in NE Asia: prerequisites, conditions and ways* 186–190

Johnstone, C..M., Nielsen, K., Lewis, K., Sarmento, A. and Lemonis, G., 2006, European Commission coordinated action on ocean energy: A European platform for sharing technical information and research outcomes in wave and tidal energy systems: *Renewable Energy* 31, 2, 191–196.

Jones, A.T. and Westwood, A., 2005, Recent progress in offshore renewable energy technology development: *Proc. IEEE Gen. Meet. Power Engng Soc.* 2, 2017–2022.

Jones, A.T. and Rowley, W., 2001, Recent developments and forecasts for renewable ocean energy systems: *TTS/IEEE Oceans 2001* 1, 575–578.

Jwo-Hwu, Y., 1998, Electric power generation at the ebb tide: *Electric Power Systems Research* **6**, 31–35.

Karasev, A. B., Mitenev, V. K. and Shulman, B. S., 1996, Ecological peculiarities of the parasite fauna of cod and pollock in the vicinity of the Kislaya Inlet tidal power plant, western Murmansk (The Barents Sea): *Sarsia* **80**, 4, 307–312.

Kerr, D., 2005, Marine energy: getting power from tides and waves: *Proc. Inst. Civ. Eng -Civ. Engng* . 158, sp. iss. 2, 32–39.

Kiho, S. and Suzuki, K., 1998, Study on the power generation from tidal currents by a Darrieus turbine: *6th International Offshore and Polar Conference–Proceedings* ed. by Chung, J. S., Olagnon, M. and Kim, C.H. 97–102.

Kirby, 1997, Environmental consequences of tidal power in a hypertidal muddy regime. The Severn estuary (UK): *La Houille Blanche-Rev. Int. de l'Eau* **52**, 3, 50–58.

Kowalik, S., 2004, Tide distribution and tapping into tidal energy: *Oceanologia* 46, 3, 291–331.

Lang, C., 2003, Harnessing tidal energy takes new turn: *IEEE Spectrum* 40, 9, 13.

Leonhard, W., 2005, Possibilities of renewable energy supplies [Möglichkeiten erneubare Energie Produktion, in German] *Electricitätswirtschaft* 104, 12, 78–83.

Lewis, G., 2002, Tidal flow energy system: *Engng Technol.* 5, 10, 19–20.

Li, L., Ren Hongliang and Li, F., 2003, Application of LongWorks Fieldbus technology in tidal current electricity control system: *Control & Automation* 3, 8–9 [in Chinese].

Little, C. and Mettam, C., 2004, Rocky shore zonation in the Rance tidal power basin: *Biol. J. Linnean Soc.*51, 1 2, 169–182.

Marfenin, N. N., Malutin, O. I., Pantulin, A. N. and Pertzova, N. M., 1997, A tidal power environmental impact assessment: Case study of the Kislogubskaia experimental tidal power station (Russia): *La Houille Blanche-Revue Internationale de l'Eau* **52**, 3, 101–107.

Mauad, F.F. and Mariotoni, C.A., 2000, Reaction turbine with simple curvature blades. An option for the generation of energy in small hydroelectric plants: *Informacion Tecnologica* 11, 5, 11–16. [in Spanish].

Meisen, P.,B., 2002, There is no energy crisis: *Proc. Winter Meet. 2002-IEEE Power Engng Soc.* 1, 117.

Millborrow, D., Hartnell, G. and Cutts, N., 1998, *Renewable energy in the EU:* London, Financial Times Energy, 316 pp.

Molland, A.F., Bahaj, A.S., Chaplin, J.R. and Batten, W.M.J., 2004, Measurements and predictions of forces, pressure and cavitation on 2-D sections suitable for marine current turbines: *Proc. Inst. Mech. Eng. [Part M] J. Engng Mait. Env.* 218, M-2, 127–138.

Montalverne, G. 1991. Moinhos de maré. Atlantis 1, 91, 67–71.

Mueller, M.A. and Baker, N.J., 2005, Direct drive electrical power take-off for offshore marine energy converters: *Proc. Inst. Mech. Eng.*[Part A, J. Power and Energy] 219, A-3, 223–234.

Munk, W., 1998, Once again: once again—tidal friction: *Journal of Progress in Oceanography* **40**, 1–4, 7–35.

Nandy, A., 1997, Renovation of the bulb in the Rance tidal power plant: **52**, 3, 47–48.

Nekrasov, A. V. and Romanenkov, D. A., 1997, On effects produced by tidal power plants upon environmental conditions in adjacent sea areas: *La Houille Blan-che- Revue Internationale de l'Eau* **52**, 3, 88–97.

O'Donnell, P., 2005, Update '05: ocean wave and tidal power generation projects in San Francisco: *Proc. Gen. Meet. IEEE Power Engng Soc.*n 2, 1990–2003.

Ottewell, S., 2003, Ireland's renewable island: *Power Eng.* 17, 3, 10–11.

Papaloisou, J.C.B. and Ivanov, P.B., 2005, Oscillations of rotating bodies: a self-adjoint formalism applied to dynamic tides and tidal capture: *Monthly Notices of the Roy. Astron. Soc.*364, 1, 66–70.

Parker, D. M., 1993, Environmental implications of tidal power generation: *IEEE Proceedings—A Science Measurement and Technology* **140**, 1, 71–75.

Pearce, N., 2005, Wolrdwide tidal current energy developments and opportunities for Canada's Pacific coast: *Int. J. Green Energy* 2, 4, 365–386.

Pelc, R. and Fujita, R.M., 2002, Reneable energy from the ocean: *Marine Policy* 573, 2, 829–844.

Petejan, S., 2003, Renewable energy – Underwater wings could give tidal power a flying start: *New Scientist* 179, 2414, 20.

Retiere, C., 1994, Tidal power and the aquatic environment of La Rance: *Biological Journal of the Linnean Society* **51**, 1–2, 25–36.

Retiere, C., Bonnot-Courtois, C., LeMao, P. and Desroy, N., 1997, Ecological status of the Rance River basin after 30 years of operation of a tidal power plant: *La Houille Blanche-Rev. Int. de l'Eau* **52**, 3, 106–107.

Riom, N. I., 1997, Tidal power plants around the world: *La Houille Blanche-Rev. Int. de l'Eau* **52**, 3, 49.

Russell, E., 2004, Could the tide be turning? (tidal power generation): *Int. Power Gener.* 27, 3, 24–27.

Ryrie, S. C., 1995, An optimal contribution model of tidal power generation: *Applied Mathematical Modelling* **19**, 2, 123–126.

Salequzzaman, M. and Newman, P., 2001, Environmental impact of renewable energy: *Proc. Australasian Univers. Power Engng Conf.* [AUPEC 2001] 625–633.

Salvi, J., 1997, The Rance tidal power plant rebuilding the alternators: *La Houille Blanche-Rev. Int. de l'Eau* **52**, 3, 42–46.

Sanford, L., 2003, Winning the tidal race: *Modern Power Syst.* 23, 7, 11–12.
Scheer, H., Ghandi, M., Aiken, D., Hamakawa, Y. and Palz, W. (eds), 1994, *The Yearbook of renewable energies*: London, James & James (Sci. Publ.) 328 pp. (Note: This book is updated each year. Latest edition 2001).
Semenov, V. N., 1997, Ecosystem alterations influences by the tidal power station on the coast of the Barents Sea: *La Houille Blanche-Revue Internationale de l'Eau:* **52**, 3, 86–87.
Shaw, T. L., 1997, Study of tidal power projects in the UK, with the exception of the Severn Barrage: *La Houille Blanche-Rev. Int. de l'Eau* **52**, 3, 57–65.
Shern, J-C., Tu, Y-Y., Tsai, H-S. and Young, DE-C., 2003, Long-term observation and analysis of the tide current flow adjacent to the nuclear power plant in Northern Taiwan: *Monthly J. Taipower Engng* 656, 82–92 [in Chinese].
Sheth, S. and Shahidepour, 2005, Tidal energy in electric power systems: *Gen. Meet. IEEE power Engng Soc.* 1, 630–635.
Shiono, M., Suzuki, K. and Kiho, S., 2000, An experimental study of the characteristics of a Darrieus turbine for tidal power generation: *Electrical Engineering in Japan* **132**, 3, 38–47.
Shiono, N., Suzuki, K. and Kiho, S., 2003, Comparison of water turbine characteristics using different blades in Darrieus water turbines used for tidal current generation: *Trans. Inst. Electr. Eng. of Japan* 123-B, 1, 76–82 [in Japanese].
Siddiqui, O. and Bedard, R., 2005, Feasibility assessment of offshore wave and tidal power production: a collaborative public/private partnership: *Proc. Gen. Meet. IEEE Power Engng Soc.* 2, 2004–2010.
Smith, D., 2005, Why wave, tide and ocean current promise more than wind: *Modern Power Syst.* 25, 5, 47–53.
Smith, D.J., 2001, Big plans for ocean power hinge on funding and additional R&D: *Power Engng* 105, 11, 91–96.
Soares, C., 2002, Tidal power: The next wave of electricity: *Pollution Engng* 34, 7, 6–9.
Spare, E.I.P., 2003, Tidal power is a wash-out: *Profess. Engng* 16, 15, 24.
Stacey, M.W., 2005, Review of the partition of tidal energy in five Canadian fjords: *J. of Coastal Research* 21, 4, 731–750.
Stamberger, A., 1993, Large tidal generating stations: *Elektrotechnik* 44, 5, 39–41.
Streets, D.G., 2003, Environmental benefits of electricity grid interconnections in Northeast Asia: *Energy* 28, 8, 789–807.
Takenouchi, K., Okuma, K., Furukawa, A. and Setoguchi, T., 2006, On applicability of reciprocating flow turbines developed for wave power to tidal power conversion: *Renewable Energy* 31, 2, 209–223.
Taylor, S. J., 1998, Sustainable development in the use of energy for electricity generation: *Proceedings of the Institution of Civil Engineers. Civil Engineering:* **126**, 3, 126–132.
Trenaman, N. and Marsden, R., 2003, Horizontal accoustic Doppler current profile instrument development and applications: *Proc. IEEE/OES 7th Working Conf. on Current Measur. Techn.* 7–11.
Twidell, J., 1993, *Energy for rural and island communities*: Oxford, Pergamon, (GB) 352 pp.
Ullman, P.W., 2002a, Offshore tidal power – beyond the barrage: *Modern Power Syst.* 22, 6, 38–39.
Ullman, P.W., 2002b, Offshore tidal power generation – A new approach topower conversion of the oceans' tides: *Mar. Techn. Soc. J.* 36, 4, 16, 24.
Ullman, P.W., 2002c, Power comes in with the tide: *Int. Water Power & Dam Constr.* 54, 9, 24–27.
UNESCO (ed.), 1993, *International directory of new and renewable energy information sources and research centres* (3rd edition): London, James & James, 605 pp.
Usachev, I.N. and Vorob'ev, I.N., 2003, Modern floating power engineering structures in Russia: *Power Techn. & Engng* 37, 4, 201–206.
Usachev, I.N., Shpolyanskii, Y.B. and Istorik, B.P., 2004, Performance control of a marine power plant in the Russian Arctic coast and prospects for the wide-scale use of tidal energy: *Power Technol. & Engng* 38, 4, 188–193.
Van Walsum, E., 2003, Barriers against tidal power: *Int. Water Power & Dam Constr.* 55, 9, 34–41.

Van Walsum, E., 2003, Barriers to tidal power: environmental effects: *Int. Power Water & Dam Constr.* 55, 10, 38–43.
Wood, J., 2002, Testing times for tidal power: *Int. Water Power & Dam Constr.* 54, 12, 16–18.
Wood, J., 2004, Racing the waves (Wave and tidal power technology): *Power Eng.* 18, 3, 24–26.
Wood, J., 2005, Marine renewables face paperwork barrier (Marine power generation): [UK] *Inst. Elec. Eng. Rev.* 51, 4, 26–27.
Wrixon, G. T. *et al.*, 1993, *Renewable energy:* Heidelberg-Berlin, Springer-Verlag. 126 pp.
Yakunin, L. P. ND Tarkhova, T. I., 1998, The estimation of the tidal energy of the Okhotsk Sea for practical usage: *6th International Offshore & Polar Engineering Conference Proceedings*, ed. by Chung, J.S., Olagnon, M. and Kim, C.H. pp.103– 105.
Young, R. M., 1995, Requirements for a tidal power demonstrations scheme: *Proceedings of the Institution of Mechanical Engineers (Part A) Journal of Power and Energy* 215–220.
Zhikui, Z., 1992, Comparison of siltation protection measures in a Chinese tidal power station, in Larsen, P. and Eisenhauer, N. (eds), *Proceedings of the Fifth International Symposium on River Sedimentation (Karlsruhe, FRG)* 847–842.

# Annex II: Additional References

ABBOT IH, 1959, THEORY WING SECTION
AISIKS EG, 1993, POWER ENG REV, V13, P4
ALDRIDGE JN, 1993, J PHYS OCEANOGR, V23, P207
ALEVIZON WS, 1989, B MAR SCI, V44, P646
ALGER TW, 1975, PERFORMANCE 2 PHASE
ALLEN JRL, 1990, MAR GEOL, V95, P77
ALLEN JRL, 2000, QUATERNARY SCI REV, V19, P1155
ALVAREZ O, 1997, 4 JORN ESP ING PUERT, V2125, P125
ALVAREZ O, 1999, ESTUAR COAST SHELF S, V48, P439
Anonymous,1964, EDF31625 DIR ET RECH
Anonymous,1968, HOUILLE BLANCHE
Anonymous,1973, HOUILLE BLANCE
Anonymous,1977, REASSESSMENT FUNDY T.P.
Anonymous,1981, WAVE TIDAL ENERGY
Anonymous,1982, ETUDE SEDIMENTOLOGIQ
Anonymous,1984a, PRELIMINARY STUDY SM
Anonymous,1984b, PUBLIC POWER NOV, P99
Anonymous,1986, KOREAN TIDAL POWER S
Anonymous,1988, 1 MERS BARR CO REP
Anonymous,1989, 57 EN PAP
Anonymous [WORLD ENER COUNC], 1993, REN EN RES OPP CONST
Anonymous [DAN EN AG], 1999, WINDP DENM TECHN POL
Anonymous [GREENPEACE], 2001a, POW SEA OFFSH REV
Anonymous, 2001b, ECONOMIST MAY, P78
Anonymous, 2001c, FUTURIST, V35, P2
Anonymous [US HOUS COMM], 2001d, SCI TECHN 7 REP WAV
Anonymous [SEA SOL POW INT], 2001e, POW SUN VIA SEA
Anonymous [MAR CURR TURB LTD], 2002a, TECHN NEW MILL
Anonymous [US Dep.Of Ener.], 2002b, DOEEIA03832002
Anonymous [US DOE], 2002c, INT EN OUTL
Anonymous [SCOTTISH EXECUT., 2003, SEC REN FUT SCOTL RE
APEL JR, 1985, J PHYS OCEANOGR, V15, P1625
APPLEBY PG, 1992, URANIUM SERIES DISEQ, P731
ARAUJO IB, 2001, IN PRESS GLOBAL OCT
ARROYO JM, 2000, IEEE T POWER SYST, V15, P1098
AUBREY DG, 1985, ESTUAR COAST SHELF S, V21, P185
AVERY WH, 1994, RENEWABLE ENERGY OCE
BACHELET G, 1990, MER, V28, P199
BAETEMAN C, 1993, QUATERNARY SHORELINE
BAHAJ AS, 2003a, RENEW ENERG, V28, P2205

BAHAJ AS, 2003b, RENEW ENERG, V28, P2205
BAINES PG, 1973, DEEP-SEA RE, V20, P179
BAINES PG, 1982, DEEP-SEA RES, V29, P307
BAINES PG, 1985, DYNAMICS ATMOSPHERES, V9, P291
BAIRD S, 1993, ENERGY FACT SHEET OC
BAKER AC, 1986, TIDAL POWER
BAKER AC, 1989, POWER ENG J NOV, P345
BAKER SK, 1993, PUBLIC LIBRARY Q, V13, P3
BANAI M, 1982, ENERGY DIGEST OCT, P39
BANAL M, 1981, BHRA FLUID ENG, P327
BANAL M, 1989, HOUILLE BLANCHE, V6
BARNEA J, 1981, FUTURE SMALL ENERGY
BASSINDALE R, 1941, P BRISTOL NAT SOC, V9, P143
BASTOS AC, 2002, THESIS U SOUTHAMPTON
BEDELL SA, 1987, GEOTHERMAL RESOURCES, V16, P3
BELL TH, 1975, J GEOPHYS RES, V80, P320
BENDER E, 2001, TECHNOLOGY REV
BENDER LC, 1993, J GEOPHYS RES-OCEANS, V98, P16521
BERNSHTEIN LB, 1986, WATER POWER DAM CONS, P37
BERNSHTEIN LB, 1989a, 1ST P INT C OC EN RE, P167
BERNSHTEIN LB, 1989b, HYDROTECH CONSTR, V22, P687
BERNSHTEIN LB, 1994, POWER ENG REV, V14, P18
BERNSHTEIN LB, 1995, IEEE T ENERGY CONVER, V10, P591
BERNSHTEIN LB, 1996, TIDAL POWER PLANTS
BERNSHTEIN LB, 1997, TIDAL POWER PLANTS
BERTHOIS L, 1954, B LABORATOIRE MARITI, V40, P4
BERTHOIS L, 1955, B LABORATOIRE MARITI, V41, P3
BEUKEMA JJ, 1979, NETH J SEA RES, V13, P203
BINNIE, 2001, ETSUT0600209REP DEPT
BLACKFORD BL, 1978, J MAR RES, V36, P529
BLANTON J, 1969, J GEOPHYS RES, V74, P5460
BLANTON JO, 2000, ESTUARIES, V23, P293
BLANTON JO, 2001, ESTUARIES, V24, P467
BLOOMFIELD KK, 1999, GEOTHERMAL RESOURCE, V23, P221
BOHNSACK JA, 1989, B MAR SCI, V44, P631
BONNOTCOURTOIS C, 1997, ATLAS PERMANENT MER, V3, P29
BOORMAN LA, 1994, HYDROGRAPHIC SOC SPE, V33, P1
BOURCART J, 1957, B SOC GEOLOGIQUE FRA, V7, P345
BOUSSUGES, 1966, 9e SHF JOURN HYDR
BOYDEN CR, 1977, FIELD STUDIES, V4, P477
BRANDT P, 1999, J GEOPHYS RES-OCEANS, V104, P30039
BRONICKI L, 1989, 14th C WORLD EN C MONT
BROOK A, 2000, GAMSCPLEX 7 0 US NOT
BRYANS AG, 2005, INT COUNCIL LARGE EL
BRYDEN IG, 1994, P 1 EUR WAV EN C UK
BRYDEN IG, 1995, UNDERWATER TECHNOL, V21
BRYDEN IG, 1998, ENERGY INT J, V23
BRYDEN IG, 2002, JOR3CT980205
BRYDEN IG, 2004, WORLD REN ENG C 8 DE
BUEKER RA, OCEANS 1991 S HONOLULU
BURD F, 1996, 185 ENGL NAT
BURGER M, 2005, COMMUNICATION 0212
CAHOON DR, 2000, GEOLOGICAL SOC LONDO, V175, P223
CAMPBELL RG, 1989, GEOTHERMAL RESOURCES, V13, P565

## Annex II: Additional References

CAMPMAS P, 1961, HOUILLE BLANCHE N° de MAI
CARETTI L, 1992, P WAVE ENERGY WORKSH
CARMICHAEL AD, OCEAN ENERGY RECOVER
CARTER I, 2004, HYDROPOWER DAMS, V1, P80
CARTWRIGHT DE, 1969, NATURE, V223, P928
CARTWRIGHT DE, 1991, J GEOPHYS RES-OCEANS, V96, P16897
CAZAUX C, 1970, THESIS U BORDEAUX 1
CAZENAVE P, 1964, EXEMPLE AMENAGEMENT
CHANG GW, 2001, IEEE T POWER SYST, V16, P743
CHARNEY G, 1969, OKEANOLOGIYA, V9, P143
CHIN DA, 1999, WATER RESOURCES ENG, P14
CHOW VT, 1959, OPEN CHANNEL HYDRAUL
CHRISTOFFERSEN JB, 1985, OCEAN ENG, V12, P387
CHUANG WS, 1981, J PHYS OCEANOGR, V11, P1357
CLAESON L, 1987, ENERGI FRAN HAVETS V
CLAVIER J, 1983a, 17EME ACT S EUR BIOL, P75
CLAVIER J, 1983b, B SOC SCI BRETAGNE, V55, P93
CLAVIER J, 1983c, P 17 EUR S MAR BIOL, P75
CLAVIER J, 1985, RAPPORT ETUDE ASS MI
COLLIGAN JG, 1993, DOEEIA0348
COLLIGNON J, 1991, ECOLOGIE BIOL MARINE
CONEJO AJ, 2002, IEEE T POWER SYST, V17, P1081, id. P1265
CONSTANS J, 1979, MARINE SOURCES ENERG, CH5
COUCH SJ, 2001, THESIS U STRATHCLYDE
COUCH JC, 2004, P 3 INT C MAR REN EN
COX C, 1962, J OCEANOGR SOC JPN, V20, P499
CRAIG PD, 1987, J MAR RES, V45, P83
CRAIG PD, 1994, J PHYS OCEANOGR, V24, P2546
CUMMINS PF, 1997, J PHYS OCEANOGR, V27, P762
CUNDY AB, 1995, MAR CHEM, V51, P115
CUNDY AB, 1996, ESTUAR COAST SHELF S, V43, P449
CUNDY AB, 1997, ENVIRON SCI TECHNOL, V31, P1093
DADSWELL MJ, 1994, BIOL J LINN SOC, V51, P93
DAGAN G, 1986, WATER RESOUR RES, V22, S120
DALE AC, 1996, J PHYS OCEANOGR, V26, P2305
DALE AC, 2001, J PHYS OCEANOGR, V31, P2958
DAMELIO L, 1935, IMPIEGO VAPORI ALTO
DANIEL T, SUSTAINABLE DEV INT, P121
DA SILVA JCB, 1998, J GEOPHYS RES-OCEANS, V103, P8009
DAUVIN JC, 1987, MAR ENVIRON RES, V21, P247
DAUVIN JC, 1993, OCEANIS S D, V19, P25
DAVIES AG, 1988, J GEOPHYS RES-OCEANS, V93, P491
DAVIES AM, 1991, INT J NUMER METH ENG, V12, P17
DAVIES AM, 1994a, INT J NUMER METH FL, V18, P163
DAVIES AM, 1994b, J PHYS OCEANOGR, V24, P2441
DAVIES AM, 1995, J PHYS OCEANOGR, V25, P29
DAVIES AM, 1999, AGU SERIES, V56
DEAN RG, 1986, COAST ENG, V9, P399
DECHAMPS C, 1982, IEEE T PAS, V101, P113
DEKERGARIOU G, 1971, SCI PECHES, V205, P11
DELATORRE S, 2002, IEEE T POWER SYST, V17, P1037
DESROY N, 1998, THESIS U RENNES 1 FR
DESRUISSEAUX M, 1993, UNPUB OCEANOGRAPHIE
DEYOUNG B, 1989, J PHYS OCEANOGR, V19, P246

DJORDJEVIC VD, 1978, J PHYS OCEANOGR, V8, P1016
DOLIVEIRA B, 2002, MACMILLAN REED NAUTI
DRAINVILLE G, 1968, NAT CAN, V95, P805
DRONKERS J, 1986, NETH J SEA RES, V20, P117
DUCKERS LJ, 1992, SEP WORLD REN EN C, V2, P2306, id. P2541
DUCKERS LJ, 1994, WORLD RENEWABLE ENER, V3, P1444
DUFF GFD, 1974, IRIA C
DUNBAR DS, 1987, J GEOPHYS RES-OCEANS, V92, P13075
DUSHAW BD, 1995, J PHYS OCEANOGR, V25, P631
EGBERT GD, 1994, J GEOPHYS RES, V99, P24821
EGBERT GD, 1997, PROG OCEANOGR, V40, P53
EGBERT GD, 2000, NATURE, V405, P775
ERMAKOV SA, 1992, DYNAM ATMOS OCEANS, V16, P279
FALCAO A, 1989, AJNA ISES SOLAR WORL
FALNES J, 1991, ENERG POLICY, V19, P68
FAY JA, 1981, 2ND INT S WAV TID EN, P409
FISCHER E, 1929, ANN I OCEANOGRAPHIQU, V5, P201, id., P205
FISCHERPIETTE E, 1931, ANN I OCEANOGR, V10, P213, id.215
FISCHERPIETTE E, 1933, B I OCEANOGRAPHIQUE, V619, P1
FITZGERALD J, 2004, GENERATION ADEQUACY
FLATHER RA, 1976, MEM SOC ROY SCI LI 6, V10, P141
FLAVIN C, 1997, FINANCING SOLAR ELEC, V10
FLYNN T, 1997, GEOTHERMAL RESOURCES, V26
FLYNN WW, 1968, ANAL CHIM ACTA, V43, P221
FRAENKEL PL, 1999, SUSTAINABLE DEV INT, V1, P107
FRANKEL P, 2000, POWER GENERATION REN, P221
FRAENKEL P, 2002a, EUR ENERGY VENTUREN F
FRAU JP, 1988, RANCE TIDAL POWER ST
FRAU JP, 1989, 20EME C PAR ANN UN O
FRENCH CE, 2000, CONT SHELF RES, V20, P1711
FRENCH JR, 1992, EARTH SURF PROCESSES, V17, P235
FRENCH JR, 1993, EARTH SURF PROCESSES, V18, P63
FRENCH PW, 1999, OCEAN COAST MANAGE, V42, P49
FRIEDRICHS CT, 1988, ESTUAR COAST SHELF S, V27, P521
FRIEDRICHS CT, 1992, J GEOPHYS RES-OCEANS, V97, P5637
FRONTIER S, 1976, J RECH OCEANOGR, V1, P35
FRONTIER S, 1983, STRATEGIE ECHANTILLO
FRONTIER S, 1991, ECOSYSTEMES STRUCTUR
GAGNON L, 2002, HYDROPOWER DAMS, V4, P115
GAILLARD JM, 1958, B LABORATOIRE MARITI, V44, P7
GANDON M, 1973, 6 ANS EXPLOITATION U
GARRETT C, 1984, ENDEAVOUR NEW SER, V8, P160
GARRETT C, 2002, IN PRESS CAN APPL MA
GARRETT CJR, 1977, J PHYS OCEANOGR, V7, P171
GASPAR P, 1990, J GEOPHYS RES-OCEANS, V95, P16179
GAUDIOSI G, 1996, RENEW ENERG, V9, P899
GAUDIOSI G, 1999, RENEW ENERG, V16, P828
GERKEMA T, 1996, J MAR RES, V54, P421
GERKEMA T, 2001, J MAR RES, V59, P227
GIBRAT R, 1976a, PONTRYAGIN CALCULS E
GIBRAT R, 1976b, SCI TECHNIQUES SEP
GILL AE, 1982, ATMOSPHERE OCEAN DYN
GILLIBRAND PA, 1995, J PHYS OCEANOGR, V25, P1488
GLORIOSO PD, 1997, PROG OCEANOGR, V40, P263

## Annex II: Additional References

GLOWKA D, 1997, GEOTHERMAL RESOURCES, V21, P405
GODIN G, 1988, TIDES
GOLDBERG ED, 1963, GEOCHRONOLOGY 210PB, P121
GOLDWAG E, 1989, TIDAL ENERGY
GORBAN AN, 2001, J ENERG RESOUR-ASME, V123, P311
GORDON DC, 1979, MARINE POLLUTION B, V10, P47
GORLOV AM, 2001, ENCY OCEAN SCI, P2955
GOWER JFR, 1980, NATURE, V288, P157
GRALL JR, 1972, THESIS U PARIS
GRANT AD, 1981, 2 INT S WAV TID OTEC, P117
GRANT WD, 1979, J GEOPHYS RES, V84, P1797
GRATTON Y, 1994, FJORD SAGUENAY MILIE, P8
GRAY AJ, 1990, SPARTINA ANGLICA RES, P5
GREEN MO, 1990, J GEOPHYS RES-OCEANS, V95, P9629
GREENBERG DA, 1979, MAR GEOD, V2, P161
GREHAN A, 1991, OPHELIA S, V5, P321
GREMARE A, 1998, J SEA RES, V40, P281
GROSSMAN GD, 1997, FISHERIES, V22, P17
GUIXIANG L, 1991, COLLECTION OCEANOGRA, V14, P128
GUTIERREZ JM, 1996, GEOGACETA, V21, P155
HAIDVOGEL DB, 1986, J PHYS OCEANOGR, V16, P2159
HAMMOND NW, 1989, DEV TIDAL ENERGY
HAMMONS TJ, 1993, P IEEE, V89, P419
HARLEMAN DRF, 1966, ESTUARY COASTLINE HY, P493
HARVEY J, 1985, DEEP-SEA RES, V32, P675
HASSELMANN K, 1988, J PHYS OCEANOGR, V18, P1775
HAWS ET, 1997, P I CIVIL ENG-WATER, V124, P1
HAWS ET, 1999, P I CIVIL ENG-WATER, V136, P117
HAYDOCK JL, 1982, TID POW C BEDF I OC
HEALEY RG, 1981, ESTUARINE COASTAL SH, V13, P535
HEALY M, 1962, PROGR SOIL ZOOLOGY, P3
HEAPS N, 1981, P 7 BARR S LOND, P35
HEATHERSHAW AD, 1979, GEOPHYS J ROY ASTRON, V58, P395
HEATHERSHAW AD, 1981, MAR GEOL, V42, P75
HEATHERSHAW AD, 1985, CONT SHELF RES, V4, P485
HENDY NI, 1964, FISHERIES INVESTIGAT, V1
HENNAGIR TL, 1997, PHILIPPINES FAST TRA
HENRY RF, 2001, MAR GEOD, V2493, P139
HERPE E, 1986, SOC SCI BRETAGNE, V58, P19
HERVOUET JM, 2000, HYDROL PROCESS, V14, P2209
HERZOG SJ, 1999, APPRAISAL J, V67, P24
HEYDT GT, 1993, P IEEE, V81, P409
HILL AE, 1998, J MAR RES, V56, P87
HILLARIET P, 1984, HOUILLE BLANCHE, V8, P571
HLINAK AJ, 1982, GRC C SMALL SCAL GEO
HOGANS WE, 1985, AM SCI
HOGANS WE, 1987, AM SCI
HOLLIGAN PM, 1985, NATURE, V314, P348
HOLLOWAY G, 1987a, J FLUID MECH, V184, P463
HOLLOWAY PE, 1987b, J GEOPHYS RES-OCEANS, V92, P5405
HOLLOWAY PE, 1999, J GEOPHYS RES-OCEANS, V104, P25937
HOLMES C, 1955, THESIS U MAINE BANGO
HOOKER JN, 1986, INTERFACES, V16, P75
HOWARTH MJ, 1990a, OTH89293 DEPT EN

HOWARTH MJ, 1990b, UK DEP ENERGY PUBLIC, V89, P293
HUBBARD JCE, 1965, J ECOL, V53, P799
HUBBERT MK, 1971, SCI AM, V225, P60
HUTHNANCE JM, 1981, PROGR OCEANOGRAPHY, V10, P193
HUTHNANCE JM, 1984, J PHYS OCEANOGR, V14, P795
HUTHNANCE JM, 1986a, COASTAL ESTUARINE SC, V3, P1
HUTHNANCE JM, 1986b, P ROY SOC EDINB B, V88, P83
HUTHNANCE JM, 1987, PROGR OCEANOGRAPHY, V19, P177
HUTHNANCE JM, 1993, PHILOS T ROY SOC A, V343, P569
HUTHNANCE JM, 1995, PROGR OCEANOGRAPHY, V35, P353
IBANEZ F, 1988, MAR ECOL-PROG SER, V49, P65
IBANEZ F, 1993, CR ACAD SCI III–VIE, V316, P745
ILIC M, 1998, POWER SYSTEMS RESTRU
INALL ME, 2000, J GEOPHYS RES-OCEANS, V105, P8687
INALL ME, 2001, CONT SHELF RES, V21, P1449
JACKSON S, 2004, P HYDR 2004 NEW ER H
JAMES ID, 1980, ESTUARINE COASTAL MA, V10, P597
JEANS DRG, 1998, THESIS U WALES BANGO
JEANS DRG, 2001, CONT SHELF RES, V21, P1855
JEANS DRG, 2001, J MAR RES, V59, P327
JOHNS B, 1975, J GEOPHYS RES, V80, P5109
JOHNS B, 1978, J PHYS OCEANOGR, V8, P1042
JOHNS B, 1987, 3 DIMENSIONAL COASTA, P17
JOINT I, 2001, DEEP-SEA RES PT II, V48, P3049
JONES AT, 2002, OCEANS ENERGY
JONES B, 1993, U932 UCES U WAL
JONSSON IG, 1966, P 10 INT C COAST ENG, P127
JONSSON IG, 1980, OCEAN ENG, V46, P75
KAGAN BA, 1972, IZV AS ATMOS OCEAN P, V8, P533
KAGAN BA, 1997, PROG OCEANOGR, V40, P109
KAGAN BA, 1999, IN PRESS IZVESTIYA A
KANTHA LH, 1995, J GEOPHYS RES-OCEANS, V100, P25283
KANTHA LH, 1997, PROG OCEANOGR, V40, P163
KANTHA LH, 2000, NUMERICAL MODELS OCE
KARASEV AB, 1996, SARSIA, V80, P307
KARMARKAR N, 1984, COMBINATORICA, V4, P373
KECK H, 2005, P IND HYDR 2005 NEW, P74
KEEN TR, 1995, J PHYS OCEANOGR, V25, P391
KELLEY JE, 1960, J SOC IND APPLIED MA, V8, P703
KEMP PH, 1982, J FLUID MECH, V116, P227
KENNISH MJ, 1998, POLLUTION IMPACTS MA, P80
KENYON KE, 1969, J GEOPHYS RES, V74, P6991
KERR D, 1989, DEV TIDAL ENERGY
KESTIN J, 1980, SOURCEBOOK PRODUCTIO
KIDDER LE, 1995, PHYS REV D, V52, P821
KIHO S, 1991, ESC918 IEEJ NEW EN C, P67
KIHO S, 1992, T IEE JPN D, V112
KIHO S, 1998, 6th INT OFFSH POL C P, P97
KILLWORTH PD, 1978, J PHYS OCEANOGR, V8, P188
KIRBY R, 2005, P I CIV ENG ENG SUST, V158, P31
KIRKE B, 2003, UNPUB DEV DUCTED WAT, P12
KIRKEGAARD JB, 1978, MEDDR DANM FISK HAVU, V7, P447
KONDRATYEV KY, 1969, RAD ATMOSPHERE
KONYAEV KV, 1995, DEEP-SEA RES PT I, V42, P2075

## Annex II: Additional References

KOWALIK Z, 1993a, J GEOPHYS RES-OCEANS, V98, P16449
KOWALIK Z, 1993b, NUMERICAL MODELING O
KOWALIK Z, 1998, J PHYS OCEANOGR, V28, P1389
KRIEGER Z, 1984, 4578953, US
KUNDU P, 1990, FLUID DYNAMICS
KWONG SCM, 1997, PROG OCEANOGR, V39, P205
LACAZE JC, 1976, B MUSEUM NATIONAL HI, V386, P71
LAKKOJU VNMR, 1996, RENEW ENERG, V9, P870
LAMAO P, 1985, THESIS RENNES FRANCE
LARGE WG, 1981, J PHYS OCEANOGR, V11, P324
LARGE WG, 1982, J PHYS OCEANOGR, V12, P464
LARGIER JL, 1990, CONT SHELF RES, V10, P759
LARSEN PF, 1984, 35 BIG LAB OC SCI TE
LAUGHLIN G, 2001, ASTROPHYS J 2, V551, L109
LAUGHTON MA, 1990, 22 WATT COMM REP
LAVELLE JW, 1991, J GEOPHYS RES-OCEANS, V96, P16779
LAWTON FL, 1972, TIDAL POWER, P105
LEBLOND PH, 1978, WAVES OCEAN
LEE DS, 1989, 1ST P INT C OC EN RE, P150
LEE STY, 1978, IEEE T POWER APPAR S, V97, P1769
LEIBOWITZ HM, 1990, GEOTHERMAL RESOURC 2, V14, P1037
LEMAO P, 1986a, J MAR BIOL ASSOC UK, V66, P391
LEMAO P, 1986b, OISEAU REV FRANÇAISE, V56, P171
LEPROVOST C, 1985, J PHYS OCEANOGR, V15, P1009
LEPROVOST C, 1994, J GEOPHYS RES, V99, P24777
LEPROVOST C, 1997, PROG OCEANOGR, V40, P37
LEVITUS S, 1982, NOAA PROF PAP, V13, P173
LEVITUS S, 1994a, WORLD OCEAN ATLAS 19, V3, P99
LEVITUS S, 1994b, WORLD OCEAN ATLAS 19, V4, P117
LI CY, 1997, J GEOPHYS RES-OCEANS, V102, P27915
LITTLE C, 1980, ESTUARINE COASTAL MA, V11, P651
LIU AK, 1985, J PHYS OCEANOGR, V15, P1613
LONG AJ, 1995, J COASTAL RES, V17, P299
LONG AJ, 1999, QUATERN INT, V55, P3
LONG AJ, 2000, GEOLOGICAL SOC LONDO, V175, P253
LONGUETHIGGINS MS, 1968a, J FLUID MECH, V31, P417
LONGUETHIGGINS MS, 1968b, J FLUID MECH, V34, P49
LOU J, 1997, ESTUAR COAST SHELF S, V45, P1
LOUCKS RH, 1975, REPORT PHYSICAL OCEA
LOZIER MS, 1995, PROGR OCEANOGRAPHY, V36, P1
LUND JW, 2001, GEOTHERMICS, V30, P29
LYLE S, 1991, NUMERICAL METHODS OP
MACARTHUR RH, 1967, THEORY ISLAND BIOGEO
MACLEOD A, 2002, P MAREC 2002
MADDOCK L, 1978, ESTUARINE COASTAL MA, V6, P353
MANNING R, 1890, T INSTN CIVIL ENG, V20
MARCHENKO A, 1999, EUR J MECH B-FLUID, V18, P511
MARCHUK GI, 1977, OCEAN TIDES
MARCY GW, 2001, ASTROPHYS J 1, V556, P296
MARDLING RA, 2002, UNPUB APJ
MARIE P, 1938, B LABORATOIRE MARITI, V20, P68
MARSDEN RF, 1994, J PHYS OCEANOGR, V24, P1097
MARTIN WR, 1970, J FISH RES BOARD CAN, V17, P169
MATHIVATLALLIER H, 1989, CAH BIOL MAR, V30, P473

MCCAVE IN, 2001, DEEP-SEA RES PT II, V48, P3107
MELLOR GL, 1982, REV GEOPHYS SPACE PH, V20, P851
MELLOR GL, 1996, J PHYS OCEANOGR, V26, P2214
MELVILLE GT, 2001, P MAREC 2001
MELVILLE WK, 1994, J PHYS OCEANOGR, V24, P2041
METTAM C, 1979, MARINE POLLUTION B, V10, P133
METTAM C, 1983, HYDROBIOLOGIA, V99, P155
MICHALLET H, 1998, J FLUID MECH, V366, P159
MIDDLETON N, 2001, GEOGRAPHICAL, V73, P52
MIGAZAKI T, OCEANS 91 S HONOLULU
MILLER P, 1996, REMOTE SENSING ACTIV, P265
MIYAZAKI T, 1991, 3RD S OC WAV EN UT J
MOROZOV EG, 1995, DEEP-SEA RES PT I, V42, P135
MOSONYI E, 1963, WATER POWER DEV, V1, CH4
MUFFLER LJP, 1979, 790 US GEOL SURV
MULKAY EL, 1998, J MECH DESIGN, V120, P17
MUNK W, 1997, PROG OCEANOGR, V40, P7
MUNK W, 1998, DEEP-SEA RES PT I, V45, P1977
MURATA S, 1982, T JSME A, V48
MURRAY CD, 1999, SOLAR SYSTEM DYNAMIC
NAEF D, 2001, ASTRON ASTROPHYS, V375, L27
NEKRASOV AV, 1990, ENERGY OCEAN TIDES
NEKRASOV AV, 1992, J KOREAN SOC COAST O, V4, P168
NELSON AR, 1996, J GEOPHYS RES-SOL EA, V101, P6115
NELSONSMITH AJ, 1967, FIELD STUDIES, V2, P407
NEW AL, 1988, DEEP-SEA RES, V35, P691
NEW AL, 1990, DEEP-SEA RES, V37, P513, id.,P1783
NEW AL, 1992, DEEP-SEA RES, V39, P1521
NEW AL, 1999, OCEAN SUBSURFACE LAY, P173
NEW AL, 2000, PROG OCEANOGR, V45, P1
NIET TA, 2001, THESIS U VICTORIA BC, P126
NIGOGHOSSIAN JP, 1973, OPTIMIZATION DESIGN, P346
NILSSON O, 1998, IEEE T POWER SYST, V13, P959
NORRIS S, 1986, 8 MAFF
ODONNELL P, 2002, COMMUNICATION
OKUBO A, 1971, DEEP-SEA RES, V18, P789
OLIVE PJW, 1977, J MAR BIOL ASSOC UK, V57, P133
OLIVER FW, 1920, ANN APPL BIOL, V7, P25
OLIVIER F, 1996a, J EXP MAR BIOL ECOL, V199, P89
OLIVIER F, 1996b, J SEA RES, V36, P217
OLIVIER F, 1997, THESIS U RENNES 1 FR
OSBORNE P, 1998, ELECT SEA
OSTROVSKY LA, 1989, REV GEOPHYS, V27, P293
PADMAN L, 1992, J GEOPHYS RES-OCEANS, V97, P12639
PALMA ED, IN PRESS J GEOPHYS R
PARKER BB, 1991, TIDAL HYDRODYNAMICS, P237
PARKER DM, 1993, IEE PROC-A, V140, P71
PATCHEN RC, 1981, 4 NOS OC CIRC SURV
PAULSON CA, 1977, J PHYS OCEANOGR, V7, P952
PEALE SJ, 1999, ANNU REV ASTRON ASTR, V37, P533
PEARSON TH, 1985, J EXP MAR BIOL ECOL, V92, P157
PEIN A, 1912, GEZEITENWASSERKRAFT
PELLETIER RR, 1972, SEDIMENTATION PATTER, P153
PENTCHEFF D, 2004, WWW TIDE CURRENT PRE

# Annex II: Additional References

PERCIVAL SM, 2001, ASSESSMENT EFFECTS O
PERKINS H, 1994, J PHYS OCEANOGR, V24, P721
PETERSON T, 2005, COMMUNICATION 0218
PETHICK JS, 1981, J SEDIMENT PETROL, V51, P571
PIANKA ER, 1970, AM NAT, V104, P592
PIESOLD D, 2002, P I CIVIL ENG-CIV EN, V150, P124
PINGREE RD, 1973, J PHYS OCEANOGR, V3, P280
PINGREE RD, 1979a, J MAR BIOL ASSOC UK, V59, P689
PINGREE RD, 1979b, MAR GEOL, V32, P269
PINGREE RD, 1982, CONT SHELF RES, V1, P99
PINGREE RD, 1983, J MAR BIOL ASSOC UK, V64, P99
PINGREE RD, 1984a, J MAR BIOL ASSOC UK, V64, P889
PINGREE RD, 1984b, REMOTE SENSING SHELF, P287
PINGREE RD, 1985, PROG OCEANOGR, V14, P431
PINGREE RD, 1986, NATURE, V321, P154
PINGREE RD, 1988, SMALL SCALE TURBULEN, V46, P387
PINGREE RD, 1989a, DEEP-SEA RES, V36, P735
PINGREE RD, 1989b, PROG OCEANOGR, V23, P303
PINGREE RD, 1990, J MAR BIOL ASSOC UK, V70, P857
PINGREE RD, 1991, J PHYS OCEANOGR, V21, P28
PINGREE RD, 1992a, DEEP-SEA RES, V39, P1147
PINGREE RD, 1992b, J GEOPHYS RES-OCEANS, V97, P14353
PINGREE RD, 1993, DEEP-SEA RES PT II, V40, P369
PINGREE RD, 1994, J MAR BIOL ASSOC UK, V74, P107
PINGREE RD, 1995a, DEEP-SEA RES PT I, V42, P245
PINGREE RD, 1995b, J MAR BIOL ASSOC UK, V75, P235
PINGREE RD, 1999, CONT SHELF RES, V19, P929
POLALRD H, 1966, MATH INTRO CELESTIAL
POLLARD RT, 1973, GEOPHYS FLUID DYN, V3, P381
POLLARD RT, 1996, PROG OCEANOGR, V37, P167
POLZIN KL, 1997, SCIENCE, V276, P93
PONTES MT, 2001, P 18 WORLD EN C BUEN
POSTMA H, 1961, NETHERRLANDS J SEA R, V1, P148
PRANDLE D, 1981, 2ND INT S WAV TID EN, P397
PRANDLE D, 1984a, ADV WATER RESOUR, V7, P12
PRANDLE D, 1984b, ADV WATER RESOUR, V7, P21
PRANDLE D, 1984c, ADV WATER RESOUR, V7, P21
PRANDLE D, 1984d, PHILOS T ROY SOC A, V310, P407
PRICE R, 1991, NOV WORLD CLEAN EN C, P1
PUGH DT, 1987a, HDB ENG SCI
PUGH DT, 1987b, TIDES SURGES MEAN SE
QUINN T, 1990, ASTRON J, V99, P1016
RAINE R, 1998, CONT SHELF RES, V18, P883
RAMESH R, 1997, RENEWABLE ENERGY TEC
RANDERSON PF, 1979, ESTUARINE COASTAL LA, P48
RAO SS, 1996, ENG OPTIMIZATION THE, CH7
RAY RD, 1996, GEOPHYS RES LETT, V23, P2101
RAY RD, 1997, PROG OCEANOGR, V40, P135
REED MJ, 1983, 892 US GEOL SURV
REES AP, 1999, DEEP-SEA RES PT I, V46, P483
REISE K, 1987, HELGOLANDER MEERESUN, V41, P69
RETIERE C, 1976, CR HEBD ACAD SCI, V282, P1553
RETIERE C, 1979, THESIS U RENNES 1 FR
RETIERE C, 1980, CONSEQUENCES ECOLOGI

RETIERE C, 1984, HOUILLE BLANCHE, V8, P583
RETIERE C, 1986, 20e C SAINTM ANN E
RETIERE C, 1989, HOUILLE BLANCHE, V2, P133
RITCHIE JC, 1990, J ENVIRON QUAL, V19, P215
ROBBINS JA, 1978, BIOGEOCHEMISTRY LEAD, P285
ROBINSON AWH, 1955, GEOGR J, V121, P33
RONEY JA, 1984, GUIDE OCEAN THERMAL
ROSA LP, 1996, ENVIRON CONSERV, P2
ROUSE M, 1950, ENERGIE MAREES
ROUVILLOIS A, 1967, CAH OCEANOGR, V19, P375
ROUX D, 2004, P HYDR 2004 NEW ER H
ROUZE M, 1959, ENERGIE MAREES
SALTER SH, 1998, P 3 EUR WAV POW C
SANDQUIST EL, 2002, ASTROPHYS J 1, V572, P1012
SANDRIN P, 1980, EDF B DIRECTION E B3
SAULSON SH, 1988, GEOTHERMAL RESOURCES, V12, P423
SCHAFER CT, 1990, OCEANOGRAPHY LARGE S, P378
SCHRICKE V, 1982, SAUVAGINE CHASSE, V2219, P16
SCHWIDERSKI EW, 1979, 79414 NSWC TR
SCHWIDERSKI EW, 1981, TR3866 NSWCDL TR
SEATON M, 2003, P I CIVIL ENG-CIV EN, V156, P124
SEIBERT GH, 1979, J FISH RES BOARD CAN, V36, P42
SEMENOV VN, 1997, HOUILLE BLANCHE, V52, P86
SERPETTE A, 1989, CONT SHELF RES, V9, P795
SHAHIDEHPOUR M, 2002, MARKET OPERATIONS EL
SHAPIRO GI, 1997, J PHYS OCEANOGR, V27, P2381
SHAW TL, 1980, ENV APPRAISAL TIDAL
SHEBLE GB, 1999, COMPUTATIONAL AUCTIO
SHERWIN TJ, 1987, ADV UNDERW TECHNOL O, V12, P263
SHERWIN TJ, 1990, DEEP-SEA RES, V37, P1595
SHIMIZU Y, 1990, APPL NATURAL ENERGY, P90
SHIONO M, 1996, ESC961 IEEJ NEW EN C, P37
SHIONO M, 2000, ELECTR ENG JPN, V132, P38
SHUM CK, 1997, J GEOPHYS RES-OCEANS, V102, P25173
SIGNELL RP, 1990, J GEOPHYS RES-OCEANS, V95, P9671
SJOBERG B, 1992, DEEP-SEA RES, V39, P269
SLAGSTAD D, 1987, STF48F87013 SINTEF
SMITH JD, 1977, SEA, V6, P539
SMITH NPA, 1992, WORLD RENEWABLE ENER, V2
SMITH WHF, 1997, SCIENCE, V277, P1956
SORENSON B, 2000, RENEWABLE ENERGY ITS
SOULSBY RL, 1987, J HYDRAUL RES, V25, P341
SOUTHWARD AJ, 1978, J FISH RES BOARD CAN, V35, P682
SOUTHWOOD TRE, 1977, J ANIM ECOL, V46, P337
SPALART PR, 1989, TURBULENT SHEAR FLOW, V6, P417
SPAULDING ML, 1987, 3 DIMENSIONAL MODELS, P405
SPEER PE, 1985, ESTUAR COAST SHELF S, V21, P207
STACEY MW, 1984, J PHYS OCEANOGR, V14, P1105
STACEY MW, 1985, J PHYS OCEANOGR, V15, P1652
STACEY MW, 1986, J PHYS OCEANOGR, V16, P1062
STACEY MW, 1992, ATMOS OCEAN, V30, P383
STACEY MW, 1995, J PHYS OCEANOGR, V25, P1037
STACEY MW, 1997, J PHYS OCEANOGR, V27, P2081
STEERS JA, 1964, COASTLINE ENGLAND WA

STIGEBRANDT A, 1976, J PHYS OCEANOGR, V6, P486
STIGEBRANDT A, 1980, ESTUAR COAST MAR SCI, V11, P151
STIGEBRANDT A, 1989a, J PHYS OCEANOGR, V19, P917
STIGEBRANDT A, 1999b, J PHYS OCEANOGR, V29, P191
STONE R, 2003, SCIENCE, V299, P399
STOTHERS R, 2005, E COMMUNICATION 0302
SUMER BM, 1987, ADV TURBULENCE, P402
SWALLOW JC, 1977, 1977C32 HYDR COMM IN
TABOR H, 1961, UN C NEW SOURC EN RO
TAKAHASHI P, 1996, OCEAN THERMAL ENERGY
TAKAMATSU Y, 1983, ENERGY NAT RESOUR, V4
TAKAMI S, 1984, ENERGY NAT RESOUR, V5
TANG YM, 1996, J GEOPHYS RES-OCEANS, V101, P22705
TAYLOR GB, 1975, DATA SERIES BEDFORD
TAYLOR GI, 1921, P LOND MATH SOC, V20, P148
TAYLOR JR, 1993, DYNAM ATMOS OCEANS, V19, P233
TETT P, 1984, OCEANOGR MAR BIOL, V22, P99
THERRIAULT JC, 1984, ESTUAR COAST SHELF S, V19, P51
THOMPSON DR, 1986, NATURE, V320, P345
THOMSEN L, 1998, PROG OCEANOGR, V42, P61
THORPE SA, 1987a, J FLUID MECH, V178, P279
THORPE SA, 1987b, PHILOS T ROY SOC A, V323, P471
THORPE SA, 1988, DEEP-SEA RES, V35, P1665
THORPE SA, 1990, PHILOS T ROY SOC A, V331, P183
THORPE SA, 1992a, P ROY SOC LOND A MAT, V439, P115
THORPE TW, 1992b, ETSU R72
THORPE TW, 1999, R120 ETSU
TINIS SW, 1995, THESIS U BRIT COLUMB
TOOLE JM, 1994, SCIENCE, V264, P1120
TRAPP T, 2001, MAREC 2001 C ENG BUS
TRAPP T, 2002a, M0201301ADT UK
TRAPP T, 2002b, P MAREC 2002
TRAPP T, 2002c, STINGRAY PROGRAMME 2
TROELSSMITH J, 1955, DANMARKS GEOL UNDE 4, V3, P1
TSUCHIYA M, 1992, DEEP-SEA RES, V39, P1885
USACHEV IN, 1988, GIDROTEKNICHESKOE ST, V12, P31
USHIYAMA I, 1971, COMPACT WIND TURBINE, P59
VAGER BG, 1969, ATMOSPHERIC OCEANIC, V5, P168
VANAKEN HM, 1996, PROG OCEANOGR, V38, P297
VANAKEN HM, 2000, DEEP-SEA RES PT I, V47, P757
VANDENBERGHE L, 1996, SIAM REV, V38, P49
VANDERBEI RJ, 1994, ORSA J COMPUTING, V6, P32
VANDERBEI RJ, 1995, PROGRAM STAT OPERATI
VANDERPLAATS GN, 1984, NUMERICAL OPTIMIZATI, CH6
VANDERWERFF H, 1958, DIATOMEENFLORA NEDER
VANGRIESHEIM A, 1990, PROG OCEANOGR, V24, P103
VLASENKO VI, 1992, PHYS OCEANOGR, V3, P417
VOS PC, 1988, GEOL MIJNBOUW, V67, P31
VOS PC, 1993, HYDROBIOLOGIA, V269, P285
WALSH JJ, 1988, CONTINENTAL SHELF RE, V8, P435
WANG X, 1992, WHOI9240
WANG YP, 1999, J COASTAL RES, V15, P471
WHITE DE, 1975, 726 US GEOL SURV
WHITE FM, 1979, FLUID MECH

WHITE M, 1997a, ANN GEOPHYS-ATM HYDR, V15, P1076
WHITE M, 1997b, SLOPE CURRENT DYNAMI, P65
WHITE PRS, 1991, NOV IMECHE M WAV EN
WHITHAM GB, 1974, LINEAR NONLINEAR WAV
WHITTAKER T, 2003, POTENTIAL MARINE CUR
WHITTAKER TJT, 1991, ETSU WV1680 CONTR RE
WILLIAMS AA, 1989, SCI TECHNOLOGY DEV, V7, P98
WILSON SS, 1977, INT J MECH ENG ED, V5
WOLFE MH, 1990, 25TH INT EN CONV ENG, P1
WORTMAN AJ, 1983, INTRO WIND TURBINE E
WRIGHT M, 2004, WAVE TIDAL TECHNOLOG
WRIGHT SJ, 1995, PRIMAL DUAL INTERIOR
WUNSCH C, 1975, REV GEOPHYSICS SPACE, V13, P167
XING JX, 1998, J GEOPHYS RES-OCEANS, V103, P27821
XUEMIN C, 1985, WATER POWER DAM CONS, P33
YAKUNIN LP, 1998, 6 INT OFFSH POL ENG, P103
YUN FC, 1958, ALL CHIN C TID POW U
ZUTIAN GE, 1989, OCEAN ENERGY RECOVER, P157

# Annex III: Special References for Chapter 2

Azurmendi, P.L. (1985) *Molinos de mar*, 71pp, Colegio Oficial de Arquitectos de Cantabria, Santander.
Azurmendi, P.L. et al. (1988) *Exposicion etnografica "Molinos de Mar". El aprovechamiento tradicional de las mareas*, 16pp, Diputacion Regional de Cantabria & Universidad de Cantabria, Santander.
Boithias, J.L. (1988) La marée motrice. Architecture des moulins à marée en Bretagne du Nord. *Le Chasse-Marée, Histoire et Ethnologie Marines* **38**, 22–37.
Boithias, J.L. & de la Vernhe, A. (1988) Les moulins à mer et les anciens meuniers du littoral: mouleurs, piqueurs, porteurs et moulagers. *Métiers, Techniques et Artisans*. Créer 275pp.
Branco, F.C. (1990) *Moinhos de mar em Portugal*: Universidade da Lisboa, Lisboa.
Charlier, R.H. (1982) *Tidal energy*. Van Nostrand-Reinhold, New York pp. 52–74.
Charlier, R.H. (1993) *Ocean energies*. Elsevier, Amsterdam, pp. 273–275.
Cordon, J. (1975) *Molino de agua salada San Juan*. Soc. Est. Vascos-Busco-Ikaskuntza, Donostia
Cornier, Y. (1990) *Les aboiteaux en Acadie: hier et aujourd'hui*. Chaire d'Etudes Acadiennes, Collection Mouvange, Moncton NB **2**, 109, 23–29.
Escaleza, J. & Villegas, A. (1985) *Molinos y panaderias* tradicionales.Universidad de Cantabria, Santander.
Foucher, F. (1998) Les moulins à marée. *Le Chasse-Marée/ArMen* Numéro Spécial Juillet 24–25.
Guillet, J. (1982) Meuniers et moulins à marée duMorbihan. *Le Chasse-Marée – Histoire et Ethnologie marines* **5**, 42–57.
Holt, R. (1988) *The mills of medieval England*. Basil Blackwell, London.
Homualk de Lille, C. (1987) *Les moulins de l'ouest*. Vieux Chouan, Fromentine (France), pp. 24–32.
Le Nail, B. (s.d.) *Les moulins à marée de Bretagne*. L'auteur, Brest, 26pp.
Lopez, B.B. (1991) *Muinos de mares e de lento en Galicia*. Fund. P.Barrie de la Meza, La Coruña, p. 95.
Minchinton, W.E. (1977–1982) Tidemills of England and Wales. *Transactions 4th International Symposium of the Molinological Society*, 339–353.
Perez, L.A. (1985) Molinos de mar. *Coleccion. Official Arquitectos de Cantabria*, Santander pp. 13–14.
Rivals, C. (1973) Moulins à marée en France. *Transactions 3rd International Symposium of the. Molinological Society*, 320–348.
Terribas, B.(1992) Molinos de marea. De la mar al grano: *Revista MOPT [Ministerio de Obras Publicas y Transportes]* **394**, 41–45.
Triggs, A. (1989) *The windmills of Hampshire*. Ensign Publ., Southampton.
Veyrin, P. (1936) Les moulins à marée du Pays Basque. *Bulletin du Musée du Pays Basque* 414–423.

Wailes, R. (1941) Tide mills in England and Wales. *Junior Institution of Engineers & Record of Transactions* **51**, 91–114.

Wailes, R. (1961) *Tide mills*. Society for the Protection of Ancient Buildings, London.

Zu-Tian, Ge (1961) The development of tidal resources in China: the tidal power experimental station of Jiangxca and its No.1 and 2 bi-directional tidal water turbines. In Krock, H.-J., ed. *Ocean energy recovery ICOER 89* [International Conference on Ocean Energy Resources], *Proceedings of the International Conference*, 157–166.

# Annex IV: Update 2008

## Chapter 1

Besides an occasional paper in the specialized technological and/or professional literature, mostly indexed in *Oceanic Abstracts*, updating and review papers covering the entire area of ocean extracted power, can be found in the trade periodical *World Energy Review*, edited by Jacky Jones, published in Great Britain by James & James. A more academic approach is followed by the Elsevier journal (Elsevier Science Publishers, Great Britain), *Renewable and Sustainable Energy Reviews*, edited by L. Kazmeski of the National Energy Research Laboratories (Colorado, USA).

The total world resource for river hydro-plants, tidal and wave was estimated, in 1995, at 6,000 GW by Duckers. Of these, 277 Twh of tidal power are found in the former Soviet Union, 20 in Argentina (San José Gulf), 17 each in the Severn River location (United Kingdom) and Turnagain Arm (Alaska). The Gulf of Cambay (India) and Cobequid Bay (Canada) could provide respectively 15 and 14 Twh. Bernshtein and Usachev (1997) reminded, on the occasion of the 30th anniversary of the TPP at Kislaya Guba, that Russia has a 17 million kW capacity and could transfer to the combined European power system 50 Twh per year and the proposed Tugur station could add a capacity of 8 million Twh/year utilisable by the coastal regions of the Russian Federation and Japan. He pointed out that if the American proposal of constructing a combination transport and power tunnel under the Behring Strait would ever materialize, a Penzinskaya TPP could be part of the complex with a capacity of 87 million kW. The potential of China is commonly overlooked, even though it attains 110 million kW. Five hundred bays in seven provinces share this resource: Tchekian, Shantoung, Kouandtong, , Iangsou, Hopei and Kianing. Duckers reviewed low head hydro-electric schemes and drew parallels with tidal power plants.

Among proposed innovations for TP schemes the HVDC[1] bus "comprising transmission lines and converter sub-stations" could be used to "connect tidal power sources and tie interconnected IEPS[2] of the Russian Far East and Siberia" (Belyaev 2002, 2003).

---

[1] High voltage direct current.
[2] Interconnected Electric Power Systems.

In China the technology of the LonWorks Fieldbus digital communication network has been applied in the 70 kW tidal current electricity generating system. It was thus shown that local devices and working conditions can be managed from an office and that working processes can be logged and adjusted in time. Incidentally the Fieldbus links intelligent local devices and automatic system with multi-directional structures (Li 2003).

A valuable companion paper to Charlier's "The view from China" appeared in the same periodical. Chang et al. (2003) review all aspects of the energy scene including tidal and tidal flow energy. The paucity of information about Chinese TPPs has been mentioned. Description have often appeared in rather obscure journals, and, naturally, frequently Chinese. The Kianghsia facility, however, was briefly described in the [French] *Revue de l'Énergie*. The first Chinese TPP is far less important than the Rance River TPP (240,000 kW) with six 500 kW turbines that will deliver 10,740 kWh annually (Song 1981).

The environmental benefits of harnessing tidal power, among such other sources as hydro-electricity and extracting natural gas in the Russian Federation's Far Eastern region, were pointed out (Streets 2003). Neighbouring areas are "choking" in the fumes emanating from coal-fired and bio-fueled electricity plants; cities in China, Mongolia, both Koreas and Japan would find present relief, not even mentioning considerable future benefits.

Corrosion concerns (Faral 1973; Leborgne 1973; Duhoux 1973) and cathodic protection (Legrand 1973) have been studied since the plant was barely five years on line.

The AGRA program was put in use at the Rance TPP in 1970. The AGRA new operational model for the Rance, was introduced in 1982. (Merlin 1982; Mourier et al. 1990; Nandy 1997; Salvi 1997; Sandrin 1980). Its special features include close optimization by dynamic programming, modular program structure, and developed graphical outputs.

Tidal current is still on the front stage. More than 20,000 Gwh per year could be generated in the near future on the West Coast of Canada by harnessing tidal current energy. Pearce (2005) assessed tidal current power on the basis of potential energy delivered, economic viability and environmental considerations. In his very recent paper he examined resource characteristics, conversion efficiency, sites, capacity factors, power densities, hydrodynamic feedbacks, and integration into existing grids.

Tidal stream [current] has also been considered in Ireland. Besides assessing the resource, estimated to have a mean output of 130 MW in several sites, the Bryans studies (2004, 2005) looked at methods of deployment and control, inclusive of down rating of the generator in regards turbine size and operational output reduction so as to reduce capital costs, capacity factor increase and reduction of impact on the grid. Two tidal energy devices involved horizontal-axis turbines and one had an undulating wing design. Tidal power has been investigated at Bull's Mouth: Achill, the largest Irish island, could be electrically self-sufficient using winds and tides as energy sources provided the network be upgraded (Ottewell 2003).

Relatively little research has been conducted to determine the characteristics of turbines running in water to convert kinetic energy. The fundamental issues likely to play important roles in MCT systems implementation are the particularly harsh marine environments—though already examined in the case of Kislaya Bay TPP—cavitations phenomena, and high stresses (Bajah & Myers 2003). L. Myers and A.S. Bahaj (2005, Simulated electrical power potential harnessed by marine current turbine arrays in the Alderney Race: *Renewable Energy* 30, 11, 1713–1731) submitted the design of a horizontal axis turbine; they used tidal data from Alderney Race to run simulations to show a potential annual energy output of 1340 Gwh. Such "races" exist also near Scotland's west coast. Land mass constrictions in these geographical locations are at the origin of high current velocities at depths suitable for the placement of turbines. In a preceding paper (2004) the same authors had emphasized that the power density—for marine currents—for a horizontal axis turbine is similar, in form, to that of a wind turbine and depends on the cube of the speed and the fluid density (water ds = 1000 × air ds). A tidal current turbine can thus be smaller than the wind turbine. Energy yields exceeding 7.4 TWh could—theoretically—be realized, which would satisfy 2% of the demand of the United Kingdom (2000).

Marine Current Ltd used a single turbine variant of one originally developed by IT Power. Two axial flow rotors with diameters 17–23 m approximately in diameter drive a generator via a gearbox, mounted on a tubular steel monopile about 2.7–3.3 m in diameter, set into a hole drilled in the seabed (Sanford 2003).

Environmental effects have been again re-examined by Van Walsum in 2003. The interest of that study lies in the fact that it addressed two-basin schemes, rarely still considered today, even though such scheme had been consistently considered for the Bay of Fundy and nearby sites such as Quoddy.

Gorlov of Boston's Northeastern University has been active for decades in developing systems for capturing tidal energy. A few decades ago he proposed a « removable barrage » scheme and energy storage with his pneumatic chambers. He recently (2003) introduced a helic[oid]al turbine which could be used in harnessing tidal as well as wave energy. The scheme dispenses with dams and can operate in free or ultra-low headwater currents. The cross-flow unidirectional rotation machine is usable in reversible tidal streams and estuaries, for instance.

In line with the philosophy that even small quantities of power are worth harnessing, Bryden et al. (2000) proposed to link tidal current turbines with energy storage mechanisms even if small. They furthermore (2004a,b) developed a simple model of tidal current energy extraction. He showed that flow in a simple channel is altered. Ten per cent of the flux ought to be the limit to assess the extractable amount. Admittedly the limitation may be less stringent where sea lochs are concerned.

Blunden and Bahaj (2006) used a two-dimensional tide driven hydrodynamic model to assess the Portland Bill (Dorset, GB) proposed site for a tidal current [stream] TPP. . The results could eventually be used if a TPP were indeed to be implanted there. The site being closer to population centers, it is economically attractive.

The capability of tidal currents to provide firm power was defended recently, stressing concommittantly the finite nature of wind power expansion (Clarke 2006).

Tidal current tapping must take into consideration that mutiplication of the number of turbines will not only have an unfavorable environmental effect but will also block the flow and thereby reduce power generated. The latter is substantially less than the average kinetic energy flux observed. According to Garrett (2005, 2004) the maximum average power lies between 20 and 24% of the peak tidal pressure head (from one end of the channel to the other) in the undisturbed channel. Garrett also showed that an array of turbines placed at the entrance of a bay, gulf, inlet, is quite effective if uniformly distributed across the entrance[3]. The continuous operation in a bay's entrance of current turbines is not less productive than accumulating the high tidewaters in a retaining basin for release at outgoing tide, and power generation is compatible with flushing and purposes other than power generation.

Tide-generated coastal currents are part of the nearshore ocean resources that governments are eyeing to provide relief from global warming. (Jones 2005).

Studies of tidal current harnessing were launched as collaborative public/private project early in 2005. The EPRI (Electric Power Research Institute) plays a leading role in the project (Siddiqui 2005). The promised technical-economic results for tidal current power plants and several sites have not yet been disclosed. A keen interest has also been manifested by San Francisco for a hydro-venturi approach which contrary to the Rance River one, would involved no moving parts placed underwater; the value of Blue Energy or Verdant Power vertical or horizontal axis-type propellers might be determinant in any implementation here. Conventional "air driven" turbines are to be placed on land and generate the electricity from the venturi suction[4]. Indeed, this system causes the least disturbance for the fauna of the Bay area (Hammons 2005, O'Donnell 2005, *Proc. IEEE Conf. 2005*). Arlington (Virginia) based Verdant Power is a major cog in the project to place tidal current turbines in New York City's East River (Grad 2004).

San Francisco city officials passed a resolution committing the city to the development of a 1 MW TPP during the 2003–2006 span. But 2006 is gone, and no further news concerning this proposal has been received.

Conversion of marine currents' kinetic energy, different from tidal currents', has been proposed for many decades, witness the gigantic Coriolis Project. The technology is similar to that for wind turbines but we are still quite far from a cost-effective large-scale system. Marine Currents Turbine Ltd [London, UK] has been involved in the development of such system for some time (Fraenkel 2002). Marine Current

---

[3] The maximum power available from a quadratic drag law occurs when when the tidal range inside the bay is reduced to 0.74 of the original amplitude.

[4] In a venturi tube the velocity is increased and the pressure lowered for a fluid that is conveyed through it, by a constricted passage. Named after the Italian G.B. Venturi [1746–1822].

Turbines Ltd [MCT] (Basingstoke, UK) participated in the Seaflow Project whose aim was the development of a tidal power plant based on the windmill concept. Such a scheme was established off the Devon Coast in the English Channel (Lang 2003). A 130 Mt unit was cemented to the sea-floor at about 1.1 km from the shore, protruding a few meters above sea-level. "Tidal currents turn the eleven-meter-long rotor, but as they reverse direction, the rotor's blades can be pitched to accept flow from the opposite direction.[...] The rotor turns slowly in water, [but] at 17 rpm the speed is sufficient, with appropriate gearing, to harness the tide's energy and drive a turbine". Rotor speed varying, a power-conditioning system (with Ac-Dc-Ac conversion) allows to get a current output of 50 Hz (grid frequency). Mean power reaches 100 kW with a peak of 300 kW, fulfilling the needs of about 500 homes.

Tidal power installations have been traditionally conceived as coastal project. Offshore tidal power harnessing has been suggested by Ullman (2002). The installation would involve a self-contained impoundment structure. Likely sites include southwest Alaska, Puerto Mont in Chile and Swansea Bay in the U.K.

The specific materials requirements for a tidal power plant have been discussed by Gooch (2000) with a view on capital costs reduction, harsh environment longevity prolongation, generation efficiency increase.

In turbines with two types of impulse, a wave-energy Wells turbine and a cross-flow Darrieus type one for very-low head hydropower (in a unidirectional steady flow obtained by physical tests models) were compared for their characteristics. This was done in non-dimensional forms; power plant performance with the various turbines, numerically simulated on equivalent scaled turbines; the comparison was based on the low similitude on turbine performance with the non-dimensional characteristics under one of the simplest controls in combination with a suitable reservoir ponds area. "The conclusion was reached that the öutput of the plant depends [evidently] on the tidal [range] and a pond inundation area."

If climate change and the high cost of oil are powerful incentives to search for or develop other methods to generate power, the possible exhaustion of that fossil fuel is an equally strong motive, albeit on the longer term. Assuming the current consumption of oil continues at the same rate, pessimists place the doomsday date in 2040, optimists in 2080, but all seem to agree that the supplies will be down to naught before the end of the current century. There has been talk, decades ago, to exploit the non-conventional oil reserves, viz. getting the oil out of tar-sands and oil-shale, but the plans were set aside because the technology then at hand would provide oil at a far from competitive price.

Among recent developments not related to utilization of tides, bacteria are holding some promise. Among substitute power sources is a bacteria able to produce hydrogen from sugar-saturated wastes, as proven by a test using wastes from a candy (nougat) and beverages (caramel) factory (see p. 28).

If these efforts are worthy of consideration, is it then so far-fetched to reactivate the idea of tidal energy harnessing which has a proven technology and nearly half a century of "loyal service"?

**Table 1** Non-comprehensive list of tidal power generation devices

| DEVICE | CHARACTERISTICS |
| --- | --- |
| Tide mill | Ancestral scheme including a retaining basin, sluices, wheel, occasionally Sea mill using run of river in an estuary, mostly up and down movement due to tide Most derelict or dismantled. Some reconstructed and resumed activity as tourist attraction or as part of a "working museum". |
| Tidal Power Plant[TPP] | Single or multiple basins system. Single or double (ebb and flood) effect. With or without pumping mode. Requires a barrage or dam. A model with "removal" dam has been designed. Uses bulb, straflo, or even other turbines. |
| Tidal stream Tide current | Use the tide current (horizontal movement) instead of up-and-down tidal movement. Avoid the need of a dam. Some operate in shallow waters, other schemes in deep fast moving channels. |
| Pulse stream | Developers BMT and IT . Converter uses pair of hydrofoils oscillating across tidal flow, permitting extraction of tidal energy from shallow water. Currently building 100 kW prototype. |
| Gorlov | Aleksandr Gorlov (Northeastern University, Boston) has developed a tidal power scheme with a helicoidal turbine. Needs no dam. Support from US Dept of Energy. |
| E.ON & 8 MW | project on UK west coast Uses Rotech Tidal Turbines Lunar Energy horizontal-axis turbines mounted offshore on sea-bed. Tide current system to be tested in 1 MW version at European Marine Energy Center (Orkney). |
| | [EMEC]. Minimal environmental impact (Robert Gordon University). Time frame: 2008-2010. |
| Neptune | underwater generator converting tide current energy for feeding into the grid. A central tower is anchored to seabed, it has 2 arms each supporting a 3-blade rotor. Time frame 2007. |
| Open hydro | US company associated with Alderney Renewable Energy Holdings. 250 kW device to be tested at EMEC. |
| SeaFlow | Property of Marine Current Turbines. Single rotor 300 kW device tried out (2003) off Devonshire coast. Forerunner of SeaGen. |
| SeaGen | 2 rotor device generating 3 times as much power as SeaFlow. (1 MW) Cooperative undertaking EDF, UK Gov., Marine Current Turbines. Time frame 2007. Site: No. Ireland. |
| Lynmouth | 10 MW tidal farm in Bristol Channel. An array of 12 SeaGen-s. Environmental |
| SeaGen Arrays | Impacts currently under scrutiny. Considered currently at research stage only |
| Tidel | Twin turbo device floating in the tide current, able to turn as the tide itself turns, enabling it to face the flow. Needs no support structure and offering easy installation and maintenance as it is simply moored to the seafloor. Time frame: 2008-2010 for a 1 MW for a pilot trial. Current trials held at NaRec (New and Renewable Research Centre, Newcastle) |

Annex IV: Update 2008    213

# Chapter 4

## *Canadian TPP: Annapolis-Royal*

Canada and the provinces of New Brunswick and Nova Scotia forged ahead with building the Annapolis-Royal pilot TPP. The Bay of Fundy Tidal Power Review Board conducted a survey of 23 different sites and concluded the mouths of Cumberland Bay and of Cobequid and Shepody bays provided the best sites for a TPP. It elected to lower substantially construction costs by floating in caisson modules for the powerhouse and sluiceway sections. Power produced would be economically integrated for the—then planned—generation system of the Maritime Provinces.

Though the smallest, the Cumberland Basin location it revealed itself the largest and most economical site for which system benefits were at least 90% derived from within the Maritime Integrated System. A Cumberland TPP would allow an assessment of TPP location in the Bay of Fundy, while placing the entire scheme within Canadian jurisdiction.

The option retained was that of a single basin, single-effect ebb-flow power generation. This was a departure from the system selected for the Rance and Kislaya stations. Another move away from the "models" concerned the turbines: the Canadian planners had selected the bulb-type turbines, able to function as pumps, but instead installed Straflo fixed-blades ones. The claim was made at decision time that the Straflo turbine meant substantial savings in civil works, generator and ancillary electrical equipment costs, and also provides higher performance and easier maintenance.

### The Site

Originally called Port Royal by the French settlers was a harbor on an island that was to become Nova Scotia (Nouvelle Ecosse). The French garrison was defeated by British forces who renamed the town Annapolis in honor of Queen Anne. The defensive works have been fully restored and are about 1 $1/2$ km from Hog's Island where the TPP has been built. On the Granville Ferry side a rock-fill dam measuring 225 m and 60–18 m wide from bottom to crest prevents the flooding of agricultural marshland when the tide comes in.

On the Annapolis side a control structure, with sluice gate, discharges the river flow. In short, a set-up, on the lower reaches of the Annapolis River, quite suitable to implant a TPP. Hog's Island itself is an 8 hectare isle against which dam and structure abut.

The scheme uses a reverse turbine operation mode: at high tide water is allowed to enter the retaining basin, upstream from Hog's Island, by passing through sluices and turbine; at low tide, as long as a head of at least 1.4 m is maintained, the sluices' gates are closed, and water can flow back from basin to sea, making its way through the turbines and generating electricity for approximately three hours. Thereafter the unit comes to a standstill for about 3 $1/2$ h. The basin to sea discharge involves about 360 $m^3$/s. The tides here average 6.4 m, but spring tides reach 8.7 m. Neap tides have a range of only 4.4 m. These compare with average tides in the Bay of Fundy of from 9 to 11 m, spring tides of 14.5–16 m, and neap tides of 6–7 m.

Under normal conditions a mean efficiency of 84.5% is reached; for heads between 5 and 6 m the output is 17.8 MW, but drops to 7.5 MW for a head of 2.5 m. When exceptional conditions develop—discharges exceeding 385 m$^3$/s—efficiency may be as high as 89% and output may soar to nearly 20 MW. The powerhouse is located on Hog's Island and is linked to the Nova Scotia electrical grid at Tequille by a 69 kW 5 km long transmission line. Intake and discharge canals are associated with the 46.5 m long powerhouse; the intake opening is a 15.5 m sided square that becomes circular at the entrance of distributor support casing, but is rectangular shaped at the exit measuring 14 × 11.1 m

## Bulb or Rim?

The statement, made at the time of the Rance River TPP construction, that the bulb turbine is to be credited with making harnessing of tidal energy possible, is perhaps an exaggeration. True, however, its reversible blades allow generation at both ebb and flood tide. Currently that feature tends to be omitted from new plans. Already years ago Eric Wilson, a UK authority on and enthusiast of tidal power[5], claimed that bulb turbines were expensive and that double-effect generation, in TPPs, provided only very limited benefits. The straight-flow—"Straflo"—turbine with rim-type generator has been tested for and installed at the Anna[polis-Royal facility. It was not, then, something entirely new: indeed, such a generator had already been patented by Leroy Harza in 1919. Today's "Straflo" unit is a modern version of Harza's axial flow turbine. Other turbines are currently being proposed for potential future TPPs such as Gorlov's helicoidal turbine and the time-tested Darrieus turbine.

The challenger to the bulb turbine, thirty years ago, is a low head propeller turbine, placed in a horizontal passage for water, whose generator field poles are attached to a rotor rim mounted around the periphery of the propeller. The assembly turns inside a generator-stator disposed outside the water passage of the turbine. By being attached to the *rim*, a larger rotor diameter is attained, providing larger inertia and better stability. Better electrical performance can be achieved and substantially higher generator ratings can be attained. Be it said, *en passant*, that the low head characteristic is a valuable asset as it makes more sites appropriate for TPP utilization. Four stainless steel blades, a hub and a steel rotor constitute the turbine propeller.

A. Douma, of the Nova Scotia Power Corporation, underscored, in the eighties, another feature of the "Straflo" units: "The generator is capable of immersion in water. If flooded, [after hosing] down and [being] dried out, it can be put back into service; protective coatings and materials have been thoroughly investigated to ensure they are suitable for the [marine] environment." The turbine that was installed at the Annapolis-Royal TPP has a runner with a 7.6 m diameter and 18 wicket gates.

---

[5] See "General bibliography", Annex I of this volume, and p. 85.

The rated speed of 50 revolutions per minute (50 rpm) was to provide an annual generation of 50 GWh.

A Bit of History

Tide mills had been no strangers to Nova Scotia and "modern" tidal power was not a newcomer. Years before US President F.D. Roosevelt was contemplating of tidal power facility on Passamaquoddy Bay, projects had been formulated, and submitted, to build tidal power stations in the Maritime Provinces of Canada. All of them came to naught. One two-basin tidal power scheme was actually built, "next doors" in Boston Bay, early in the 20th century, but it failed to survive the competition of other means of electricity generation. The basins became choice pieces of real estate of contemporary Boston.

About two km from Hog's Island, a replica of the old Lequille tide mill has been reconstructed, a fitting memorial to the forerunner of the TPP Annapolis-Royal scheme. Canada does not seems to have any plans, in the present phase of the 21st century, to build another TPP, but interest is being shown in capturing—at least some—the energy of tide current on the West Coast, a development that has gathered enthusiasm in Europe.[6]

If the Annapolis-Royal TPP has proven feasibility of harnessing tidal energy in the Bay of Fundy, a new development has gained a foothold in Canada. The country's first free-stream tidal power project has been deployed near Race Rocks. Known as the Pearson College-EnCana-Clean Current Tidal Power Demonstration Project, this "station" will be thus located in British Columbia's provincial ecological reserve, some 10 nautical miles southwest of Victoria. It will provide the electricity for the reserve's needs.

The scheme has an installed capacity of 65 kW using a bi-directional ducted horizontal turbine with a direct drive variable speed permanent magnet generator. Major funding was provided by EnCana Environmental Innovation Fund, helped by Sustainable Development Technology Canada. No less than six firms participate in the undertaking with Clean Current Power Systems Ltd.

## What is a tide?

The tidal movement is due to the attraction exerted by celestial bodies upon the earth. The principal bodies having this effect are the sun and the moon, with the moon, being closest to the earth having the major impact. The difference between lunar and solar day poses some problems to tidal power generation. Efforts to compensate for this are at the origin of retiming schemes. These may involve storing of water, use of hydrogen or compressed air.

---

[6] see chapter "Current from the tide current".

The tidal movement is double: the rise and fall of the water level, and the horizontal displacement of the water: the incoming tide or flood current and the seawards current of ebb.

The tidal phenomenon has been described in scores of books and papers. It had puzzled scientists for millennia. Isaac Newton is credited with explaining it. Tidal dynamics have been discussed and explained, a.o. in connection with floods, *ad libitum* in the recently re-published landmark volumes by Wood. A short discussion on their causes and characteristics is provided in this chapter and the different types of tides briefly reviewed. Semi-diurnal and diurnal tides are those of interest in connection with tidal power plants.

Tidal power generation is explained so that the various types of tidal power plants can be assessed. These include single- and double-effect schemes and single- or multiple basins ones. Calculations of relevant coefficients are shown. Basic statistics and statistical parameters involved have been summarized.

: ¿Qué es una marea?

El movimiento de la marea se origina debido a la atracción ejercida por cuerpos celestiales sobre la tierra. Los principales cuerpos que tienen este efecto son el sol y la luna; esta última que es la más cercana a la tierra, es la que tiene el mayor impacto. La diferencia entre el día lunar y solar ha representado algunos problemas a la generación del poder de la marea. Los esfuerzos para compensar dicho problema deben involucrar almacenamiento de agua, uno de hidrógeno y aire comprimido.

El movimiento de la marea es doble: el levantamiento y la caída del nivel del agua, y el desplazamiento horizontal del agua: la corriente que sube o inundación y la corriente que baja.

El fenómeno de la marea ha sido descrito en diversos libros y artículos. Ha sido un enigma para científicos durante milenios. Isaac Newton ha sido uno de los tantos que han explicado dicho fenómeno. Éste ha sido discutido y explicado en conexión con inundaciones, ad limitum, en las recientes re-publicaciones de volúmenes de referencia por Wood. Los diferentes tipos de mareas brevemente revisadas y una discusión corta de sus causas y características es aportada en este capítulo.

La generación del poder de la marea es explicada de tal manera que los diversos tipos de poder de mareas pueda ser valorada. Estos incluyen un efecto sencillo y doble en esquemas. Calculaciones de coeficientes relevantes es mostrado. Las estadísticas básicas y parámetros básicos envueltos han sido sumariados.

*La corriente desde la corriente de la marea* 4.

La energía ha sido utilizada por el hombre para proveer poder mecánico durante milenos, tal vez mucho antes que la era de la corriente; pero es particularmente desde el siglo XI y XII cuando las máquinas de marea contribuyeron al progreso industrial. Dichas máquinas hicieron uso del movimiento horizontal y vertical.

En el siglo XX y XXI, en los esfuerzos para que la marea genere electricidad, el hombre se ha enfocado en tpp-s que usó el movimiento vertical y que involucró la construcción de una presa. La presa, dique o barrera, ha sido un impedimento financiero a la construcción de más tpp-s del tamaño del Rance, o incluso hasta de los rusos o canadienses para dicha materia. En lugar de intervenir con ligeros golpes el levantamiento y caída de mareas, es concebible usar la corriente marea. Ambas corrientes pueden ser usadas. La barrera no es necesaria.

De esta manera, diversos esquemas actualmente propuestos se encuentran en una moderna transposición de carrera de generadores de mareas. Estas pueden ser construidas en una fracción del costo de una barrera, pero la cantidad de poder obtenido es más modesto. Las estaciones pilotos han sido tratadas en varias ubicaciones geográficas y tal vez el mejor conocido es el Río del Este en Nueva Cork; actualmente bajo construcción.

*Mejoras y progresos*

Las grandes zancadas han sido hechas en construcción, operación y emplazamiento del poder de la marea desde las estaciones de Rance, Kislaya y Fundyn que han sido construidas hace menos de la mitad del siglo. El uso de "cofferdams", otro gasto además de la construcción de barreras, ha sido generalmente pospuesto. La turbina bulbo, un invento revolucionario en la época del tpp Rance fue concebida, ha sido suplida en algunos esquemas por la turbina Starflo y otros estilos que aún están en consideración.

La generación de poder para corrientes que bajan y aumentan, no está vista tan económicamente justificada con los extra costos involucrados, y operaciones de bombeo están bajo re-evaluaciones. El progreso en electrónicos han modificado operaciones y han permitido la reducción de personal. Nuevas protecciones anti corrosivas han sido introducidas.

Si las propuestas han sido colocadas sobre la mesa que está observando enormes plantas en varias regiones del mundo, ninguna es seriamente considerada, excepto para una planta mayor en la República de Corea. El uso de particularidades geomorfológicos, tales como penínsulas, han generado proyedctos que pueden liderar la implementación. Una cierta tendencia a construir pequeñas plantas, incluso reintroduciendo el generador de marea, ha sido anunciada con un ojo de poder de producción en regiones remotamente pobladas donde la electricidad central sería difícil de situar. Mientras que el ttp ha sido considerado, en el peor de los casos, como algo que tiene un impacto benigno en el ambiente, lo que ha puesto en duda la precisión de los estudios de evaluación en la materia.

*Rance, Kislaya, Fundy*

La planta de poder de marea del río Rance celebró su cuarenta aniversario en el 2006. La planta de la bahía de Kislaya está a punto de alcanzar la misma edad en un año más; al igual que la isla Hog y la bahía de Fundy. Todas han realizado un buen trabajo, incluso si ninguno ha inflamado el torbellino de construcción de plantas grandes. Se supone que hay alrededor de cientos de pequeñas plantas y unas cuantas grandes tpp-s en China, un país con un enorme potencial de poder de mareas.

El capítulo provee una descripción de las plantas y una discusión de las modificaciones y mejoras hechas alrededor de las últimas cuatro décadas.

## Chapter 9: New Developments

Since 1999, the United Kingdom has committed some £25 million on precompetitive research and development for wave and tidal stream testing facilities, besides another £5 million for EMEC [European Marine Energy Center, Orkney, Scotland] wave and tidal [stream] testing facilities. The Marine Renewable Development Fund received "state aid" approval from the European Commission.

The Seaflow Project is a tidal current (tidal stream) technology operated by Marine Current Turbines. It receives support from the DTI's technology program and the European Commission. It was launched in 2003 and is the forerunner of SeaGen, and developed simultaneously with such other tidal power devices as TidEL's stream generator (developer is Soil Machine Dynamic) and Rotech Tidal Turbine (developer is Lunar Energy).

Waves' and tidal stream's consenting arrangements for England and Wales were published in 2005 by TDI; they apply only pre-commercial energy devices) (see text at www.dti.gov.uk/renewables/publications/pdfs/guidanceonconsentingarrangements.pdf.)

The United States have finally stepped into the practical, if experimental, stage in New York City. With the Roosevelt Island Project (RITE) five areas of environmental impact were retained. They are fish movement and protection, navigation and security, recreational resources, historical resources and water quality. Fish protection is the most extensive observation topic encompassing monitoring before and after turbines installation, mobile surveys and netting throughout the study timetable. On the operational side were retained total turbines' and generators' efficiency. Finally the Kinetic Power Hydro System (KHPS) has six key characteristics among which a very low rotation per minute speed (32 rpm) for 5 m diameter turbines, open and wide spacing (12–30 m apart), very small footprint in the channel (1518 m$^2$). The actually protected area sets aside 5700 m$^2$.

Russell, among others[7], unequivocally states that the benefits of tidal flow have been largely ignored. Indeed, compared to other sources of renewable energy, tidal flow is reliable, predictable, of known gravity and sustainable. It is furthermore acceptable from an aesthetic viewpoint, being out of sight, and, contrary to wind power, creates no noise.[8] In San Francisco a sub-sea tidal power station has been under consideration for some time; no moving parts are involved, except on land, advantage would be taken of a venturi, and conventional air-driven turbines would be installed.

It should be more often stressed that tide generated electricity has no fuel costs, wastes, hazardous by-products, or emissions.

If maintenance is a flaw in the system, there are probably ways to solve this problem, and it does not justify the low profile it has been kept in. Cathodic corrosion protection, another "warning" often uttered has been improved, witness a study at

---

[7] Anonymous, 2004, Plugging into the ocean: *Machine Design Int.:* 76, 18, 82–90.

[8] Russell, E., 2004, Could the tide be turning? : *Int. Pow. Gener. [UK]* 27, 3, 24–27; Neil, J., Hassard, J. and Jones, A.T., 2003, Tidal power for San Francisco: *Oceans 2004*, 1, 7–9.

**Table 2** Examples of tidal energy developers and installations from around the world, by country and research institution or corporate enterprise

**Australia**
Tidal Energy Pty Limited (tidal energy), Queensland, Australia

**Belgium**
ESHA, European Small Hydropower Association, Brussels
Eufores, European Forum for Renewable Energy Sources, Brussels
HAECON, Harbour Engineering Consultants, Drongen-Ghent

**Brazil**
Leony Polatti-Tidal Energy

**Canada**
Blue Energy Canada (turbines for current, tidal, OWC energy), Alberta

**China**
China New Energy (tidal energy, ocean current, wave energy, thermal energy, salinity gradient energy)
University of Shanghai Jiad Tong School of Naval Architecture & Ocean Engineering 1945 Huashan Rd, Shanghai CN-200030

**France**
Comité de Liaison des Energies Renouvelables, Montreuil sur mer
Electricité de France (Tidal barrage), La Rance River, St Malo, Britanny
IFREMER, Institut Français de Recherches sur la Mer, Paris
International Energy Agency, Paris
Sogreah (Société Grenobloise d'Applications Hydrauliques), Grenoble

**Germany**
Deutsche Wasserstoff Verband, Berlin
Institut fuer Energie und Umwelt, Leipzig

**Greece**
DAEDALUS Informatics (turbines for current, tidal, OWC energy), Athens
CRES, Center for Renewable Energy Sources, Pikermi, Athens

**Japan**
Kairen Co Ltd – Changs-Ascending Energy Corp., Saitama
Saga university, Tokyo
Tokyo University of Technology, Tokyo

**Korea**
Korea Water Resources Corporation (KOWACO) (tidal energy), Sihwa
Daewood Construction Company (tidal plant construction) Seoul

**Netherlands [The]**
Neptune Systems (offshore, nearshore tidal current and wave energy)
De Kleine Aarde, (consultancy), Boxtel

**Norway**
Hammerfest Stroem AS (tidal energy), Finmark
New Energy Systems AS PB 38, N-1324 Lysaker

**Russia**
Kislaya Bay Tidal Power Station, Kisgalobskaia, Murmansk

**Sweden**
Turab turbine & Regulator Services (turbines) Nässjö

**Switserland**
Escher-Wyss AG, (tidal plant turbines) Zuerich

**Table 1** (continued)

**United Kingdom**
BMT and IT Power (Pulse Stream 100), London
BMT Group Ltd Goodrich House 1 Waldegrave Rd Teddington TW11 8LZ
British Wind Energy Associati on (BWEA)
Renewable Energy House, 1 Aztec Row N1 OPW City [of London] University, London
European Marine Energy Centern (EMEC) Orkney, Scotland
Hydroventuri (marine tidal current energy), London
IT Power Ltd(marine tidal current energy), Grove House, Lutyens Close, Chineham, Hampshire RG24 8AG
Lancaster University, Renewable Energy Gp Dept of Engineering Lancaster LA1 4YR
Lupus Engineering Services Ltd Engineering Design Div. Abergeldie Cottage, Crethier, Ballater
Marine Current Turbines, Ltd. (turbines for current, tidal, OWC energy), The Court, The Green, Stoke Grifford, Bristol, BS34 6PQ Hampshire
METOC PLC Exchange House, Station Rd, Liphook, Hanks, GU30 7DW
New and Renewable Research Centre (NaRec), Newcastle
Robert Gordon University (marine tidal and current energy), Aberdeen, Scotland
School of Mechanical Engineering, University of Edinborough (tidal energy), Scotland
Seacore (tidal power), Gweek, Helston, Cornwall
Scottish and Southern Energy (Neptune)
Scottish and Southern Energy PLC Inveralmond House 200 Dunkeld Rd, Perth,
Scottishpower UK PLC Cathcart Business Park Spean Str, Glasgow G44 4BE
Scottrenewables Ltd Hillside Office, Stromness, Orkney KW16 3HS Scotland
SMD Hydrovision (Tidel Machine), Newcastle
The Engineering Business Limited (nearshore tidal energy)
Thropton Energy Services (tidal energy, power, plants) Thropton, Morpeth, Northumberland
Tidal Electric Ltd. (near shore tidal energy),West Simsbury, CN UK
Tidel hydrovision Newcastle-on-Tyne

**United States of America**
ABS Alaskan (turbines for current, tidal, OWC energy), Fairbanks, AK
GCK Technology, Inc. (Aleksandr Gorlov Helic[oid]al Turbines), San Antonio, TX
Harza Inc. (tidal power projects) Chicago, IL
Historic Tide Mills Conferences (tide mills) 5 Berkeley Lane Topsham ME
HydroVenturi (tidal energy), San Francisco, CA
Rahus Institute, Martinez, CA
Kinetic Energy Systems (tidal current energy), Ocala, FL
Tidal Electric, Inc. (nearshore tidal energy), West Simsbury, CT and Anchorage, AK
UEK Corporation (turbines for current, tidal, OWC energy), Annapolis, MD
UNITAR-Small and Alternative Hydropower, United Nations, New York, NY
Verdant Power (tidal current energy), East River, New York, NY

(Modified and excerpted from Practical Ocean Energy Management Systems, Inc., POEMS, http://www.poemsinc.org/links.html and completed by authors)

the Huaneng Dandong Power Plant (China)[9] This "huge source of energy should be elevated through research to a much higher profile. Financial requirements remain

---

[9] Liu, G-c., Wang, S-d., Song, S-j., Zu, Y-j., Sun, L., 2004, Corrosion and control of sea water circulation system of Huaneng Dandong Power Plant: *Electric Power* 37, 2, 87–88.

Annex IV: Update 2008        221

an impediment but there are also bureaucratic barriers to cope with as well.[10] To be commercially able to fly on its own wings, tidal power must overcome grid connection exigencies, licensing requirements, and not the least marketing costs. POEMS [Practical Ocean Energy Management Systems] made a broad spectrum survey that established agreement across technology and established a list of objectives that will push ocean energies, including tidal of course, towards competitive commercial viability.[11] Switching from building a wind farm in Victoria and from pine and cane use in power stations, Verdant (an Arlington VA company) would like to address commercial viability of tidal power with a plant in New York City's East River.[12]

Schemes have been investigated to convert marine energy into electricity using, instead of pneumatic, hydraulic, et al. linkages, direct drive electrical take-off. Mueller and Baker investigated a linear vernier hybrid permanent magnet machine and an air-cored tubular permanent magnet machine.[13] Two main types of generators have been introduced recently, to wit the seafloor sited and the hydroplane. Some funding is expected from the British government. Gorlov has developed a new helicoidal turbine which is "environment friendly".[14] The unidirectional rotation machine is particularly suited for reversible use in tidal streams, bays, estuaries, etc, but also in rivers and needs neither dams nor canals.

Re-Timing, Self-Timing

The Islay Energy Trust has recently approved proposals for the first commercial-sized tidal energy project. This will concern a device with 4, 5 or 6 turbines and a capacity of 2MW in Islay Sound. The island is off the Orkneys (Scotland). Costs involved hover around €10 million.

Little had been heard, for some time, about the huge Korean tidal power project. Silence was broken just before the Summer of 2008 started. Indeed a very large 11.5 m diameter units 300-turbine field has moved towards development. Geographically the site is off the Korean coast in the Wando Hoenggan Water Way. A 300 MW plant is scheduled for completion by December 2015, but the feasibility study is to be completed by July 2008 and a 1MW pilot plant put on line by March 2009. The undertaking is a joint venture of Lunar Energy—the tidal power

---

[10] Wood, J., 2005, Marine renewables face paperwork barrier: *IEE Review* 51, 4, 26–27.

[11] Anonymous, 2003, Ocen energy development. Obstacles to commercialization: *Ocean 2003* 4, 2278–2283.

[12] Grad, P., 2004, Changing tide of power generation: *Eng. Australia* 76, 11, 51–52.; Gross, R., 2004, Technologies and innovations for system change in the UK: status, prospects and system requirements of some leading renewable energy options: *Energy Policy* 32, 17, 1905–1919; Anonymous, 2003, Ocean energy development. Obstacles to commercialization: *Oceans 2003. Celebrating the past. Teaming toward the future* 4, 2278–2283.

[13] Mueller, M.A. and Baker, N.J., 2005, Direct drive electrical take-off for offshore marine energy converters: *Proc. Inst. Mech. Eng. Assn, J. Power Energy [UK]* 219, 3, 223–234.

[14] Flin, D., 2005, Sink or swim: *Power Eng.* 19, 3, 32–35; Fox, P., 2004, Designs on tidal turbines: *Power Engineer* 18, 4, 32–33; Gorlov, A.M., 2003, The helical turbine and its applications for tidal and wave power : Oceans 2003, 4, 1996.

company—Korean Midland Power—the utility—Hyundai Heavy Industries—the turbine manufacturer—and Rotech (Aberdeen. Scotland)—the designer optimizator. (Fig. 4/5)

## Chapter 8

Leaving dreams behind and getting into reality: Strangford Lough (Narrows) in Northern Ireland was the site of a world "first" when Marine Current Turbines started Sea-Gen, the first commercial-scale tidal stream turbine. It is heralded as the most environmentally friendly generation of power with zero emissions, zero noise, and close to 100% subsurface. Sonars of Tritech International will be installed on the "plant" and monitor environmental impact-such as impact on sea mammals-during the five-year long trial period.

Costs involved with the development of new, or improved, approaches to tidal energy tapping remain substantial, even when partial funded by outside sources. Some thought has been given to reduce these by using smaller scale models. Fryer and Merry[15] proposed to test the commercial viability of physical models and are implementing their idea at Wight Island's Solent Energy Centre (SOEC), once the site of numerous tide mills[16]. Five objectives are pursued: get an idea of the comparative cost of manufacturing a single device or a small number of them; assess the installation costs; obtain an indication of the relative economies of scale for mass-production; determine the relative reliability and maintenance requirements of different devices; quantify the efficiency of different devices.

The testing can be done in a towing tank or in a circulating water channel. In the first instance it is the device that is towed through standing water, and in the second the device is held in position but water is pumped through the "basin". The advantage of the larger models in towing tanks if offset, at least in part, by the extremely low initial turbulence of the stationary water.

The idea is certainly a constructive one, but one may wonder why no reference at all is made to the physical modelling done by SOGREAH in its Grenoble laboratories when the Rance River Plant was decided upon.

---

[15] Fryer, D. and Merry, S., 2008, Early-stage tidal stream generators: *Sea Technology* 49, 4, 51–55; Anonymous, 2006, *Feasibility study-Solent ocean energy centre: the case for establishing an evaluation and research centre for ocean energy technology on the Isle of Wight. Report prepared for the Isle of Wight Council:* Marine and Technical Management Consultants.

[16] Charlier, R.H., Ménanteau. Chaineux, M-C., 2004, Rise and fall of the tide mill. In: Morcos, S., Zhu, M., Charlier, R.H. et al., (eds), *Ocean sciences bridging the millennium:* Paris, UNESCO & Qingdao, PRC, China Ocean Press p. 321.

# Chapter 9

Many an ocean and/or energy scientist is puzzled by the failure of US new legislation and budget to include incentives for ocean generated power. This omission weakens the renewable energy industry at a time when *a contrario* the European Commission, and several European governments, are funding projects. Originally the bill[17] encompassed the authorization of allocating $250 million (€170 million) to the marine renewable energy industry for research, development, demonstration and commercial application to expand marine and hydrokinetic renewable energy production. It also authorized the Department of Energy to establish national marine renewable energy research and development centers, with additionally conduct a study on environmental impacts including recommendations on their mitigation.

And yet, there is apparently no lack of interest in ocean energy harnessing. Besides the parties active in that area already listed in this volume, there ought to be mentioned[18]

- Marinus Power, 440 Louisiana Street Suite 625 Houston TX 770020
- Minerals Management Service (MMS), Leonardo NJ
- National Hydropower Association (NHA), 1 Massachusetts avenue NW, Suite 852 Washington DC 20001
- Ocean Energy Council (OEC), 11985 Southern Boulevard West Palm Beach FL 33411
- Ocean Renewable Energy Coalition (OREC)

At "Energy-Ocean 2008" held in Galveston TX several aspects of tidal energy utilization were the topics of technical papers[19]:

- Blackmore, R., 2008, Tidal energy device evaluation center (TEDEC) at the Maine Marine Academy (Cianbro Corp.)
- Mackie, G., 2008, Development of a tidal energy device, the Evopodx™ (overberg Ltd).
- O'Connell, M., 2008, Siting of ocean and tidal projects (Stoel Rives).

The other sources of ocean power are not lagging behind. Wave energy is the subject of papers by Gill and Demiss at the Galveston meeting[20], Portugal has converters at work, and the West Coast, particularly Southern California is praised as having the best "wave climate" for tapping that ocean energy. Indeed narrower continental shelves are a plus as placing the devices close to shore reduce costs and minimize wave suppression by friction, the zero energy point being about $-90$ m. Load variation being unavoidable, a system must be sized for maximum efficiency

---

[17] Energy Independence and Security Act of 2007.

[18] Addition to the UK listing: Onshore Renewables and Wave & Tidal Consents

[19] Published by the conference, contact info@energyocean.com, and www.oceanenergy.com

[20] Gill, A.T., 2008, Wave energy activities in Hawaii (Hawaii Department of Business, Economics and Tourism); Denniss, T., 2008, HECO wave energy project Hawaii (Oceanlinx).

at part load, yet designed with flexibility to capture a maximum of wave energy[21]. The European Commission augmented its support for WEC systems with grants reaching several millions of euros. Information on on wave and tidal projects has been published in such reports as the *Atlas of Wave Energy Resource in Europe and Exploitation of Tidal and Marine Currents* while setting up a *European Thematic Network on Wave Energy.* Environmental impact is minimal and has been assessed recently by T.W. Thorpe.

Offshore wind harnessing has taken off with remarkable strength and the potential siting of devices on movable –floating–supports, freeing them from shallow water monopiles, will boost it further. Turbines are larger but construction and maintenance costs are considerably higher. Production, however, has a peak value usually double that of an onshore site. One may reasonably even envision combined wind and wave conversion systems combined and located in deep water.

Ocean windmill projects have been based on towers built onto the seabed near land, draw complaints they spoil the view and disturb wildlife. A StatoilHydro 2.3 megawatt windmill will be placed about six miles off the coast of Karmoey, near Stavanger on the west coast. The electricity will be sent to and through underwater cables. The windmill, with 260-foot blades, will be mounted on top of a giant spar buoy, six meters in diameter and 100 meters deep.

Held in place by three anchors with chains, the system could be used in waters as deep as 700 m. Scheduled to go on line in late 2009, a 3 m tall model has been tested in a wave tank. The experiment could prove that floating wind power is commercially viable. It could open the door to floating tidal energy capturing devices.

---

[21] Beyene, A. and Wilson, J.H., 2008, Challenges and issues of wave energy conversion: *Sea Technology* 49, 5, 43–46.

# Annex V: Companies and Organizations Involved in Tidal Power Projects, Services, and/or Research

## V.1 Equipment

ESCHER-WYSS AG, Escher-Wyssplatz, Zürich, Switserland
HARZA INC, Chicago, IL, USA
HYDROVENTURI, London, UK
NEYRPIC S.A, BP No. 75, Centre de Tri, F-38041 Grenoble, France
SEACORE, Gweek, Helston, Cornwall, TR12 6UD, Great Britain
SOGREAH, Société Grenobloise de Recherches et d'Applications Hydrauliques Grenoble, France
THROPTON ENERGY SERVICES, Physic Lane, Thropton, Morpeth, Northumberland, NE65 7HU, England
TURAB TURBINE & REGULATOR SERVICES, Förrädsgatan 2, S 57139 Nässjö, Sweden

## V.2 Services, Consultancies and Organizations♠

CITY UNIVERSITY, School of Engineering, Northampton Square, London EC1V 0HB, England
CLER, Comité de Liaison des Energies Renouvelables, 2B Rue Jules Ferry, F-93100, Montreuil, France
CRES, Center for Renewable Energy Sources, 19th Km Marathonos Avenue, GR-19009 Pikermi, Athens, Greece
DE KLEINE AARDE, Sustainable Building, POB 61, NL- 61 AD Boxtel, The Netherlands
DEUTSCHER WASSERSTOFF-VERBAND EV-DWV, Unter den Eichen 87, D-12205, Berlin, Germany
ESHA, European Small Hydropower Association, Rue du Trône 26, B-1000 Brussels, Belgium
EUFORES, European Forum for Renewable Energy Sources, European Parliament ASP 13G240, Rue Wiertz 60, B-1047 Brussels, Belgium
HAECON, Harbour & Engineering Consultants, Deinsesteenweg 101, B-9130, Drongen, Belgium
HARZA INC, Chicago, IL, USA
IFREMER, Institut Français de Recherches sur la Mer, Paris, France
INSTITUT FUR ENERGIE UND UMWELT, Institute for Energy and Environment, Torgauerstrasse 77, D-04347, Leipzig, Germany
RAHUS INSTITUTE, 1535 Center Avenue, Martinez, CA 94553, USA
UNITAR, Small and Alternative Hydropower, United Nations Plaza, United Nations, New York City, NY, USA

---

♠ Some additional companies and sources are mentioned in the text itself.

## V.3 Various Services and Products

UK BMT GROUP LTD, Goodrich House 1 Waldegrave Rd, Teddington TW11 8LZ
British Wind Energy Associati on (BWEA), Renewable Energy House, 1 Aztec Row N1 OPW
IT POWER LTD, Grove House, Lutyens Close, Chineham, Hampshire RG24 8AG, Lancaster University, Renewable Energy Gp, Dept of Engineering, Lancaster LA1 4YR
Lupus Engineering Services Ltd, Engineering Design Div., Abergeldie Cottage, Crethier, Ballater
MARINE CURRENT TURBINES, The Court, The Green, Stoke Gr fford, Bristol, BS34 6PQ
METOC PLC, Exchange House, Station Rd, Liphook, Hanks, GU30 7DW
Scottrenewables Ltd, Hillside Office, Stromness, Orkney, KW16 3HS Scotland
Scottish and Southern Energy PLC, Inveralmond House 200 Dunkeld Rd, Perth
Scottishpower UK PLC, Cathcart Business Park Spean Str, Glasgow G44 4BE
TIDEL HYDROVISION, Newcastle-on-Tyne
CHINA, University of Shanghai Jiad Tong, School of Naval Architecture & Ocean Engineering, 1945 Huashan Rd, Shanghai CN-200030
NORWAY, New Energy Systems AS, PB 38, N-1324 Lysaker
USA, The Octagon 888 Main Str, New York, NY10044

# Annex VI – Summaries

## English

### Back out of the closet and into the limelight

Charles de Gaulle has to be credited with the construction of the first large modern tidal power plant. That was some 40 years ago. Modelling and try-outs were carried out in Grenoble, France in the facilities of SOGREAH, the *Société Grenobloise d'Applications Hydrauliques,* which had been and was to be involved in several studies concerning tidal power plants.

Harnessing the energies of the tides was not a new concept, in fact it was a very old ones, as tide mills had used such energies to provide mechanical power since millennia. The Rance River was the site chosen to construct the Rance TPP, and many tide mills, still at work, had to be dismantled to make room for an electricity power plant. Plants had been considered for various geographical locations. Britain had eyed the Mersey and Severn rivers, Argentina had feasibility studies done for the San José Gulf, Australia had been considering the Kimberleys, Korea had several locations in mind such as Garolim or Inchon, the USA even started building a plant in the Passamaquoddy region and Canada examined repeatedly the Bay of Fundy.

Even Leonardo da Vinci had designed plans to use the tides as a source for power. A very abundant literature has been produced and currently international conferences centering on tapping ocean energies are frequent. We are far from the tiny slot once attributed by a benevolent conference chairperson for a couple of papers on the subject. Ocean energies, and tidal power in particular, are frequently on the menu. The number of firms involved in *ad hoc* research increases steadily, and so do pilot try-outs.

Observation, and recording, of tides goes also back to Classical Times— Herodotus and Euripides—come to mind, Newton provided an explanation.

Besides tides differences in temperatures and in salinity, waves, the strength of ocean currents—and of tidal currents, tidal streams to the British—marine winds, marine biomass, all are under consideration as renewable energy sources and seen as environmental benign. The Rance River plant transposed the age-old tide mill principles into the contemporary world. It has a barrage, a retaining pool, bulb turbines and is capable of generating electrical power both by ebb- and flood-tides. A highway linking both sides of the channel, has been placed on top. Socially and economically the facility has lifted the rather forlorn region of Brittany into the XXth and XXIst centuries.

*Poseidon to the rescue*
Not much was said at the 2002-Johannisberg "Summit" about the role the energy alternatives offered by the ocean in the coastal zone and offshore could play in stemming global warming. Realism must prevail in staggering statistics tuned down as power from the ocean, at present, cannot replace traditional sources of energy: they are complementary, but as has been pointed out, every [little] bit helps.

Use of marine winds has expanded considerably and besides huge "parks", particularly along coasts of Sweden, Denmark, The Netherlands, for instance, more modest "farms" are being established, the latest offshore Belgium and the United States. Germany, Great Britain, Ireland are not lagging much behind.

Waves are being harvested for energy. Norway and Britain are leaders in this technology but it are the Portuguese who established perhaps (2006) the largest scheme, using, it is true, British technology. The fury of storms and the opposition of tourism officials have finally yielded to the need of using non-polluting energy, the current carbon dioxide and global warming crisis being strong arguments.

Complaints of tidal power stations being capital intensive (and overlooking the longevity of such plants comparing to the far shorter life-span of traditional and even nuclear stations), and dissatisfaction of some environmental quarters with the claimed impact of barrages, have been a boost to a return to the past when tide mills used the tidal current. It is thus possible, one rediscovered, to capture tidal energy without dikes or barrages.

The matter of tides and their energy will be examined in later chapters. Be it said that at present besides the large Rance River plant, pilot installations have functioned satisfactorily in Russia, Canada and China. Modest local stations could help progress in remote regions and in less developed areas. Would the tide mill be resurrected from "it ashes" as a XXIst century Phoenix?

Words are fine, but deeds, of course, are far better. If ocean energies can be non-inexhaustible, hydrocarbons—whether from land or sea—definitely are reserves that are being depleted at a fast pace. Winds, waves, tides have been successfully tapped. Other ocean sources such as hydrogen, thermal differences and salinity have met with unequal success. Huge schemes are less probable, and among these some would create gigantic problems.

If ocean energies are especially viewed as contributors to the electricity production, applications in derived domains, such as desalination, buoys, pumps.

Of the limited number of relatively environmentally benign extraction operations that are simultaneously sustainable, harnessing the energies that the ocean dissipates have been often mentioned, frequently maligned, rarely undertaken. It is probably not preposterous to take a new look at these sources of energy which, if grandiose schemes are not proposed, may relieve the heavy burden of remote regions, poor countries, hydrocarbons deprived areas.

*Medieval engineering that lasted*
Tide mills have played an important role in the industrial development, particularly in the West. Mills dotted coasts on both sides of the Atlantic Ocean, with even some located on the Pacific. Though gradually supplanted by more advanced technologies, several remained at work well after the end of World War II. De facto they are the forerunners of today's tidal power plants. The tide mill of bygone times inspired engineers' dreams. If many sites are suitable to establish plants, their number grew considerably when low-head turbines were put on the market.

There has been no dearth of papers on the tide mill but the most remarkable books cam out in the latter part of the century authored by such aficionados de la Vernhe, Boithias, Homualk de Lille. Not much later interest for the industrial archaeology, whether in mint, reconstructed, derelict or new use condition fostered the genesis of groups set to discover, describe, restore, even re-use the old mills. While such mills can be traced back to Irak of classical times, and some were put to use by Belisarius when besieged on the Tiber River in Rome, most historians refer to the mention of sea mills in the Domesday Book. They became a quite common part of the landscape in the $12^{th}$ (England, Wales) and $13^{th}$ (Netherlands) centuries. Most were built along coasts, but some were situated on rivers (e.g. Danube), or even beneath bridges (Thames), still others made use of peninsulas or were erected on islands.. The French brought the technique to Canada, the Dutch to the region of New York and New Jersey, and others to Massachusetts and Maine.

Mills had sometimes a retaining pond, sometimes only a wheel activated by the tide current of the river. Rather sophisticated machinery was found inside the mills. The energy was mostly transformed in mechanical power, though besides grinding, some mills were used in connection with breweries, saw mills, hoisting, etc.

Mills that have and do retain presently study interest are those of Britain, France, Portugal, Spain, North America—where once more than 300 of them were at work—and, to a lesser degree, those of the Lower Countries.

There have been comments about a "revival" or "resurrection" of tide mills and their possible usefulness to provide power to remote locations and to isolated locations.

*What is a tide?*
The tidal movement is due to the attraction exerted by celestial bodies upon the earth. The principal bodies having this effect are the sun and the moon, with the moon, being closest to the earth having the major impact. The difference between lunar and solar day poses some problems to tidal power generation. Efforts to

compensate for this are at the origin of retiming schemes. These may involve storing of water, use of hydrogen or compressed air.

The tidal movement is double: the rise and fall of the water level, and the horizontal displacement of the water: the incoming tide or flood current and the seawards current of ebb.

The tidal phenomenon has been described in scores of books and papers. It had puzzled scientists for millennia. Isaac Newton is credited with explaining it. Tidal dynamics have been discussed and explained, a.o. in connection with floods, *ad libitum* in the recently re-published landmark volumes by Wood. A short discussion on their causes and characteristics is provided in this chapter and the different types of tides briefly reviewed. Semi-diurnal and diurnal tides are those of interest in connection with tidal power plants.

Tidal power generation is explained so that the various types of tidal power plants can be assessed. These include single- and double-effect schemes and single- or multiple basins ones. Calculations of relevant coefficients is shown. Basic statistics and statistical parameters involved have been summarized.

### *Rance, Kislaya, Fundy et al.*
The Rance River tidal power plant celebrated its $40^{th}$ birthday in 2006. The Kislaya Bay plant is about to reach the same age in another year, and the Hog's Island, Bay of Fundy, is close to the same age. All have performed well, even if none has ignited a construction flurry of large or even modest plants. There are supposed to be over a hundred small, and a few somewhat larger, tpp-s in China, a country with a huge tidal power potential.

The chapter provides a description of the plants and a discussion of the modifications and improvements made over the last four decades.

### *Current from the tide current.*
Tidal energy has been utilized by man to provide mechanical power for millennia, perhaps as far back as several before our current era, but it is particularly since the $11^{th}$ and $12^{th}$ centuries that tide mills contributed to industrial progress. Those mills made use sometimes of the horizontal movement of the tides, sometimes of the vertical one.

In his efforts to harness the tides to generate electricity, man has, during the $20^{th}$ and $21^{st}$ centuries focused on tpp-s that used the vertical movement and that involved construction of a dam and use of a retaining basin or impoundment area. The dam, dike or barrage, has been a financial impediment to construction of more tpp-s of the size of the Rance, or even of the Russian and Canadian ones for that matter. Instead of tapping the rise and fall of the tides, it is conceivable to use the tide current. Both ebb and flood currents can be used. No barrage is needed.

Thus, several schemes currently proposed are a modern transposition of run of the current tide mills. These can be built at a fraction of the cost of a barrage station, but the amount of power obtained is more modest. Pilot stations have been tried out in various geographical locations and perhaps the best known one is the East River (New York City) plant currently under construction.

## *Improvements and Progress*
Great strides have been made in construction, operation and siting of tidal power plants since the Rance, Kislaya and Fundy stations have been constructed less than half-a-century ago. The use of cofferdams—another big expense besides building barrages—has been generally shelved. The bulb turbine, a revolutionary invention at the time the Rance tpp was conceived, has been supplanted in some schemes by the Straflo turbine and still other types are under consideration.

Generation of power both by ebb and flood currents is not viewed as economically warranting the extra costs involved, and pumping operations are under re-assessment. Progress in electronics have modified operations and allowed reduction of personnel. New anti-corrosive protection has been introduced.

If proposals have been placed on the table that are viewing huge plants in various regions of the world, none are seriously considered, except for a major plant in the Republic of Korea. Use of geomorphologic particularities, such as peninsulas, have generated projects that could lead to implementation. A certain trend to build small plants, even reintroduce the tide mill, has been noticed with an eye of producing power in remote or lightly populated regions where electric central would be too onerous to site.

While the tpp has been considered as having at worst a benign impact on the environment, voices are heard that put in doubt the accuracy of that assessment. Studies on that matter have gained in frequency, even if these authors remain septic.

*RHC*

## *French*

Les Annexes comportent, en plus d'un "Index", une bibliographie détaillée, une liste des firmes, des institutions de recherché, des universities qui se penchent sur les questions et techniques se rapportant à l'utilisation des marées et sujets connexes. Comparée à la liste établie il y a environ trois ans, la nouvelle liste témoigne d'une augmentation du nombre de firmes atteignant approximativement 30%.

La bibliographie inclut majoritairement des ouvrages rédigés en anglais; toutefois un grand effort a été fait pour y faire figurer des travaux en français. Elle est divisée en trios parties: la première couvre les publications d'avant 1982, la seconde les années 1982-1992. La date de 1982 arbitrairement choisie correspond cependant à celle de la parution des premiers texts "modernes" (Charlier, Baker), au 15me anniversaire de la mise en service de l'usine de la Rance et à un certain nombre de modifications se rapportant à la construction et au fonctionnement des centrales marémotrices. Ceci n'empêche que des ouvrages important n'aient été publiés avant cette date (Bernshtein, Gibrat).

La troisième section réunit les publications parues depuis 1992. Quelques 22% de celles-ci portent des dates tombant dans la fourchette 1992-2007. Cette section est la seule qui, ne fut-ce qu'en nombre restreint, inclut des travaux publiés sous forme électronique; elle n'est pas aussi complète que les preceédentes car de nombreux auteurs mentionnent les ouvrages consultés en ne fournissant que des coordonnées bibliographiques incomplètes.

La Bibliographie Générale reprend la grande majorité des ouvrages mentionnés en notes infrapaginales à une exception près: pour le Chapitre 2 une bibliographie succinte a été insérée en fin de volume, constituant une Annexe III afin de faciliter toute recherche se rapportant aux moulins de marée.

Finalement, un bref glossaire constitue une annexe séparée. Les auteurs ont estimé que les lecteurs ont acquis une meilleure connaissance des termes *ad hoc* depuis leurs livres précédents (1982, 1993).

SOMMAIRE

*Qu'est-ce la marée?*
Le mouvement de marée est dû à l'attraction exercée par des corps célestes sur la terre, principalement par le soleil et la lune; l'effet de la lune est le plus important car elle se trouve le plus proche de la terre. Le décalage entre le jour solaire et le jour lunaire provoque ceretains problèmes pour la production d'électricité par les marées. Certaines techniques et demarches sont utilisées pour compenser cette situation, telles le stockage d'eau, l'utilisation d'hydrogène ou d'air comprimé. Le "mouvement" de marée est double: des courants de flot et de jusant et le marnage.

De nombreux ouvrages décrivent et expliquent le phénomène. C'est Isaac Newton que l'on crédite d'avoir donné une explication. Parmi les ouvrages détaillés qui traitent le sujet le livre récemment réédité de Wood est remarquable. Une courte explication des causes et caractéristiques des marées et les différents types sont proposes dans ce chapitre. Les marées diurnes et semi-diurnes sont celles qui nous intéressent spécialement ici.

La production d'électricité est expliquée ce qui permet une comparaison entre les différents types de centrales. Sont inclues les stations à simple et double effet et celles à un et plusieurs basins de retention.

Annex VI: Summaries 233

Les calculs et paramètres statistiques et coefficients font partie du chapitre.

### *Le courant du courant de marée*
L'énergie des marées a été utilisée pour produire de la force mécanique depuis des millénaires, mais c'est surtout depuis les 11e et 12e siècles que les moulins à marée ont contribué au développement industriel. Ces moulins faisaient appel soit au mouvement vertical soit au mouvement horizontal de la marée.

Désireux d'utiliser l'énergie des marées pour produire de l'électricité, on s'est centré, au cours des 20 et 21e siècles sur des projets de centrales requérant des digues et bassin de retention. Les digues et barrages constituent un poste financier fort lourd dans la construction d'une usine marémotrice. Il est évidemment possible de capter l'énegie des marées en utilisant non pas le mouvement vertical mais le mouvement horizontal, tant le flot que le jusant. Dans ce cas on peut se passer de barrage.

De fait faire appel au mouvement horizontal n'est autre qu'adapter aux temps modernes la technique de jadis, lorsque des moulins de mer utilisaient les courants de marée. Si la construction d'une usine sans barrage est bien moins onéreuse il n'en reste pas moins que la production d'électricité est plus basse. Des centrales pilotes ont été mises à l'essai dans différents sites; la mieux connue est probablement celle en construction sur l'East River dans la ville de New York.

### *Evolution des connaissances*
De grands pas ont été faits dans les domaines de la construction, mode opératoire et la disponibilité de sites géographiques depuis que les centrales de la Rance, Kislaya et Fundy ont été construites il y a moins d'un demi-siècle. L'emploi de batardeaux dans la construction a été abandonné, grande dépense comme les barrages ou digues. Les turbines bulbes, célébrées comme une innovation sensationnelle il y a 50 ans, ont laissé la place, dans certains cas, aux turbines Straflo, et d'autres types sont sous considération.

La production d'électricité lors des courants de flot et de jusant n'est plus considérée comme justifiée du point de vue économique vu les dépenses supplémentaires que le double-effet requiert. D'autre part les opérations de pompage sont aussi sous examen. Les avances faites dans le domaine électronique ont amené des changements dans la démarche opératoire et permis des réductions de personnel. De nouvelles techniques anti-corrosion ont aussi été introduites.

Si des projets de construction de grandes centrales, dont certaines n'ont jamais été sérieusement considérées, ont été mis en veilleuse dans diverses résgions du monde, il n'est plus question de voir se construire des stations telle celle des Minquiers. Toutefois les Coréens ont annoncé la mise en chantier de la plus grande usine marémotrice au monde dans la région du Lac Shiwa. Il est aussi possible que des projets faisant usage de parctulariés géomorphologiques, p.ex. les peninsules, voient le jour. Une certaine tendance de construire de petites usines, ou même de réintroduire le moulin à marée, en version moderne, se manifeste aussi, surtout pour des régions éloignées ou à faible densité de population.

Les centrales marémotrices ont été considérées comme ayant une influence minimale sur l'environnement. On peut donc s'interroger sur l'exactitude de certaines opinions contemporaines qui leur attribuent un danger environnemental.

*Les usines de par le monde*
L'usine marémotrice de la Rance a célébré son 40e anniversaire en 2006. Comme lors de ses 6, 10, 15, 20 et 30 ans, plusieurs "rapports" on fait le subjet de publications. Ils sont tous cites. Charlier fit une communication sur le sujet au Renewable Resources 2006 Congress tenu à Chiba (lez Tokyo). La station de la baie de Kislaya (Russie) célébrera à son tour ses 40 ans d'ici un an. Quant à celle de la Baie de Fundy (Hog's Island, Nouvelle Ecosse, Canada) elle soufflera ses 40 bougies également très bientôt. Toutes ont fonctionné avec succès, et nonobstant aucune nouvelle centrale, fut-elle grande ou odeste n'a été implantée ailleurs.

Il est vrai que d'après des renseignements invérifiables, la Chine, pays disposant d'un potentiel gigantesque, aurait construit plus de cent petites installations faisant usage de digues déjà en place, outré quelques centrales un peu plus larges. Le chapitre donne une description des centrales et des améliorations apportées tant au mode opératoire qu'aux techniques et équipement.

*RHC*

*Dutch/Flemish*

SAMENVATTING

*Een stap terug naar de openbaarheid*
Aan Charles de Gaulle liet ongeveer 40 jaar geleden de eerste grote moderne getijdencentrale bouwen. De modellen en proefopstellingen werden gebouwd bij de firma SOGREAH uit Grenoble (Frankrijk) die in verschillende studies van getijdencentrales betrokken was en nog altijd is.

De winnen van energie uit getijden was niets nieuws, getijdenmolens werden reeds millennia lang gebruikt als energiebron. De rivier Rance werd gekozen voor de "Rance TPP" en meerdere bestaande getijdenmolens moesten worden afgebroken om plaats te maken voor de nieuwe centrale. In Engeland liet men een oog vallen op de rivieren Mersey en Severn, Argentinië voerde haalbaarheidsstudies uit voor de Golf van San José, Australië keek naar de Kimberleys, in Korea bekeek men verschillende plaatsen zoals Garolim en Inchon, in de USA werd zelfs een centrale gebouwd in het gebied van Passamaquoddy en Canada onderzocht tot dat doel herhaaldelijk de Baai van Fundy.

Annex VI: Summaries 235

We vinden zelfs bij Leonardo da Vinci plannen om getijden te gebruiken als energiebron. Er werd heel wat gepubliceerd over het gebruik van energie uit oceanen en dit is ook het thema van verschillende internationale congressen. De tijd ligt lang achter ons dat een welwillende congresvoorzitter een korte sessie wijdde aan dit onderwerp. Energie uit oceanen en getijdencentrales vinden we regelmatig terug in de programma's. Het aantal firma's dat betrokken is in het ad hoc onderzoek neemt gestadig toe en zo ook de pilootstudies.

De waarneming en het opmeten van getijden gaat terug tot de Oudheid, Herodotus en Euripides komen voor ogen, Newton gaf een theoretische verklaring.

Buiten getijden worden temperatuurgradiënten en verschillen in het zoutgehalte van zeewater, golven, zeestromingen, getijdenstromingen, zeewinden en mariene biomassa als potentiële energiebron gezien. In de getijdencentrale van de Rance werden de werkingsprincipes van de oude getijdenmolens gebruikt: een afdamming met daarachter een opvangbekken en speciale turbines die zowel bij eb als bij vloed elektriciteit kunnen opwekken. Boven op de centrale werd een moderne autoweg aangelegd die beide oevers van de Rance verbindt. Sociaal en economisch gezien heeft de getijdencentrale het eerder achtergestelde Bretagne in de XXe en XXIe eeuw geloodst.

*Poseidon komt ter hulp*
Er werden op de "Top" van Johannesburg in 2002 weinig woorden gerept over de rol van alternatieve energie die uit de oceaan kan gewonnen worden in kustzones of in open zee. We moeten realistisch zijn met onzekere statistieken omdat oceaanenergie op dit ogenblik de traditionele energiebronnen niet kan vervangen: ze is complementair, maar zoals reeds eerder gezegd, alle kleintjes helpen.

Zeewind wordt de laatste jaren in toenemende mate gebruikt als energiebron. Buiten reusachtige windmolenparken - die we bijvoorbeeld aantreffen voor de kusten van Zweden, Denemarken en Nederland – werden kleinere"parken" aangelegd in de kustzone van België en van de USA. Duitsland, Groot-Brittannië en Ierland liggen niet veel achter op de andere staten.

Golfslagenergie wordt eveneens aangewend. Noorwegen en Groot-Brittannië zijn marktleiders in deze technologie, maar het zijn de Portugezen die in 2006 het grootste project hebben opgezet, weliswaar met Britse technologie. De kracht vrijgemaakt door stormen en de tegenstand vanuit de toeristische sector gaven de doorslag in het debat om niet-verontreinigende energie te produceren; de toename van koolstofdioxide en de opwarming van de Aarde lieten de balans overslaan.

Tegenstanders zeggen dat getijdencentrales heel duur zijn (waarbij de levensduur van getijdencentrales in vergelijking met traditionele- en zelfs kerncentrales wordt over het hoofd gezien) en het ongenoegen van de "groenen" over de impact van afdammingen, hebben een duw in de rug gegeven van de opwaardering van oude getijdenmolens. Eens dat deze techniek zal doorgebroken zijn, kan getijdenenergie gewonnen worden zonder dijken en afdammingen.

In de volgende hoofdstukken worden getijden en de energie die er uit kan gewonnen worden, besproken. Het volstaat hier te vermelden dat de grote centrale aan de Rance en pilootinstallaties in Rusland, Canada en China naar behoren hebben gewerkt. Bescheidener lokale centrales kunnen in afgelegen en minder ontwikkelde gebieden vooruitgang brengen. Zal de getijdenmolen in de XXIe eeuw uit zijn as herrijzen zoals een feniks ?

"Woorden wekken, daden strekken". Oceaanenergie is onuitputbaar terwijl fossiele brandstoffen – gewonnen op land of op zee – aan een snel ritme worden opgebruikt. Wind, golven en getijden werden tot dusver met succes als energiebron aangewend. Andere bronnen zoals waterstof, temperatuurgradiënten en het zoutgehalte hebben een ongekend succes. Grote installaties zijn minder waarschijnlijk omdat ze soms even grote problemen creëren.

Oceaanenergie levert in eerste instantie een bijdrage tot de elektriciteitsproductie, maar de technieken die daarbij worden toegepast kunnen ook in aanverwante gebieden zoals ontzilting van zeewater, boeien en pomptechnieken gebruikt worden.

De winning van energie uit oceanen wordt in slechts een beperkt aantal gevallen - die elk afzonderlijk milieuvriendelijk en duurzaam zijn – toegepast. Er wordt vaak over gesproken, dikwijls in negatieve termen. Het is niet ongerijmd oceaanenergie opnieuw te bekijken zonder terug te vallen op grootse projecten doch eerder te mikken op kleinschalige toepassingen die kunnen gebruikt worden in afgelegen gebieden, arme landen en gebieden zonder fossiele energiegrondstoffen.

*Middeleeuwse ingenieurstechnieken met moderne toepassingen*
Getijdenmolens hebben een belangrijke rol gespeeld in de industriële ontwikkeling, vooral in het Westen. Men kon ze aan de kusten van beide kanten van de Atlantische Oceaan vinden alsook aan de Stille Oceaan. Hoewel ze geleidelijk werden vervangen door modernere installaties, bleven sommige molens werken tot ver na de $2^e$ Wereldoorlog. Ze zijn de facto de voorlopers van de huidige getijdencentrales. De getijdenmolen van weleer sprak tot de verbeelding van menig ingenieur. Hoewel veel locaties geschikt waren voor getijdenmolens, zien we pas een aanzienlijke groei met de komst van "low-head" turbines.

Er is geen gebrek aan publicaties over getijdenmolens, maar de meest opmerkelijke boeken werden in het laatste deel van de eeuw geschreven door "aficionados" zoals de la Vernhe, Boithias, Homualk de Lille. Niet veel later was de belangstelling voor industriële archeologie van onbeschaden getijdenmolens, heropgebouwd, vervallen of nieuw, de drijfveer tot het oprichten van werkgroepen die oude molens opspoorden, beschreven, restaureerden tot zelfs hergebruikten.

Hoewel het oude Irak reeds molens kende en er werden in gebruik genomen door Belisarius op de Tiber tijdens het beleg van Rome, citeren veel historici zeemolens in het "Domesday Book". Molens werden een vertrouwd beeld in het landschap in Engeland en Wales in de 12$^e$ eeuw en in Nederland in de 13$^e$ eeuw. Ze werden vaak aan de kust gebouwd, soms aan rivieren (bv. de Donau) of zelfs onder bruggen (Theems), op schiereilanden of op eilanden. De Fransen brachten de techniek naar Canada, de Nederlanders naar het gebied van New York en anderen naar Massachusetts en Maine.

Soms hadden molens een opvangbekken, in andere gevallen was er enkel een rad dat door de getijdenstroom van een rivier werd aangedreven. In de molens werden soms ingewikkelde technieken toegepast. Energie werd meestal omgezet in mechanische arbeid die werd aangewend voor het malen, zagen, het optillen van lasten enz. Molens die een studieobject waren en nog altijd zijn, vindt men in Groot-Brittannië, Frankrijk, Portugal, Spanje, Noord Amerika – er waren daar destijds meer dan 300 in werking – en voor een beperkt deel ook in de Lage Landen.

De heropleving en de heroprichting van getijdenmolens en hun mogelijke aanwending in afgelegen gebieden en geïsoleerde levensgemeenschappen staat ter discussie.

### *Wat is een tij ?*
Getijden zijn een gevolg van de aantrekkingskracht tussen hemellichamen en de Aarde. Het gaat hier over de Zon en de Maan waarbij de invloed van deze laatste meer doorweegt door de kortere afstand tot de Aarde. De periode tussen twee opeenvolgende culminaties van Zon en Maan is niet gelijk en hiermee moet rekening gehouden worden bij de opwekking van elektriciteit in getijdencentrales bv. door water te bewaren of door tijdelijke opslag van energie in waterstof of in samengeperste lucht.Een getijdenbeweging bestaat uit twee componenten: de op- en neergaande beweging van het waterpeil en de horizontale water verplaatsing door opkomende vloed of door zeewaartse terugtrekking bij eb.

Het getijdenverschijnsel werd in tal van boeken en artikels beschreven. Eeuwenlang heeft het wetenschappers bezig gehouden. Isaac Newton gaf als eerste een wetenschappelijke verklaring. Dynamica van getijden werd besproken en uitgelegd, o.a. in relatie tot overstromingen, in het onlangs heruitgegeven werk van Woods. De oorzaken en kenmerken en de verschillende types van getijden worden kort besproken. Voor getijdencentrales zijn halfdaagse en dagelijkse getijden van belang. Opwekking van energie uit getijden wordt uitgelegd en hieruit kunnen de verschillende types van getijdencentrales afgeleid worden. Het gaat over enkelvoudige en dubbele energieopwekkingsschema's en enkele of meervoudige bassins. Terzake kenmerkende coëfficiënten worden berekend en de statistische parameters die hiervoor nodig zijn, worden samengevat.

## Electriciteit uit getijdenstroom

De mens wint al duizenden jaren mechanische energie uit getijden, maar het is slechts van in de $11^e$ en 12e eeuw dat getijdenmolens hebben bijgedragen tot industriële vooruitgang. Horizontale of vertikale waterverplaatsing werden hiervoor aangewend.

Om elektriciteit te winnen uit getijden gebruikt men in de $20^e$ en $21^e$ eeuw bij voorkeur (getijden centrales) vertikale waterverplaatsingen en hiervoor moest een afdamming en een opvangbekken of overstromingsgebied worden voorzien. Een afdamming of dijk woog in zoverre financieel door dat meergetijden centrales van het formaat van de Rance, of Russische en Canadese tegenhangers, sedertdien niet meer werden gebouwd. Er werd overgestapt naar centrales die enkel gebruik maken van getijdenstromingen bij eb en vloed en niet meer van vertikale waterverplaatsingen. Een afdamming is met deze centrales overbodig.

Verschillende op stapel staande projecten zijn in feite een moderne versie van de oude getijdenmolens. Ze kunnen gebouwd worden met een fractie van het budget van een volwaardige getijdencentrale, maar de energieproductie is bescheidener. Pilootinstallaties werden op verschillende plaatsen opgericht, en de best gekende is de centrale van East River (New York stad) die thans op stapel staat.

## Technische Vooruitgang

Getijdencentrales werden ingrijpend aangepast qua bouw, beheer en localisatie sedert de centrales aan de Rance, Kislaya en Fundy werden opgericht, minder dan en halve eeuw geleden. Kofferdammen, die net als afdammingen zware investeringen vereisten, worden niet meer gebruikt. De dubbelerichting turbine (gloeilamp vormige) – een revolutionaire uitvinding toen de Rance werd ontworpen – werd in een aantal ontwerpen door de Straflo-turbine of andere types vervangen

Energie-opwekking door zowel eb- als vloedstromingen brengt extra kosten met zich die niet opwegen tegen de opbrengst. Pompstations worden opnieuw bekeken. Nieuwe toepassingen in de elektronica hebben de werkschema's aangepast en lieten toe op personeelskosten te besparen. Tenslotte wordt ook een nieuwe anti-corrosiebescherming toegepast.

Nieuwe projecten die de constructie van grootschalige installaties met zich brachten, werden – met uitzondering van een grote centrale in de Republiek Korea – overal elders met scepsis bekeken. Projecten van getijdencentrales die gebruik maken van de geomorfologische kenmerken van het landschap, inzonderheid van schiereilanden, hebben meer kans om uiteindelijk te worden gerealiseerd. De trend om kleine centrales en zelfs getijdenmolens te bouwen, moet worden bekeken vanuit het oogpunt van energieproductie in afgelegen gebieden of in dunbevolkte zones waar een elektrische centrale economisch niet haalbaar is.

Annex VI: Summaries

Hoewel de milieu-impact van een getijden centrale als verwaarloosbaar werd beschouwd, gaan steeds meer stemmen op die de geloofwaardigheid van deze stelling in twijfel trekken. Het aantal studies waarin dit standpunt wordt vertolkt, neemt toe maar de auteurs blijven op dit punt sceptisch.

*Rance, Kislaya, Fundy e.a..*
In 2006 werd de 40$^e$ verjaardag van de ingebruikname van de Rance getijdencentrale gevierd. De centrale van de Baai van Kislaya zal een jaar later dezelfde ouderdom bereiken en die van Hog's Eiland in de Baai van Fundy, is op weg naar de 40. Deze centrales hebben naar behoren gefunctioneerd, hoewel ze geen "boom" hebben op gang gebracht in de constructie van soortgelijke eenheden. Waarschijnlijk zijn er een honderdtal meestal kleinere en enkele grotere getijdencentrales in China, een land dat een enorm potentieel heeft aan getijdenenergie.In dit hoofdstuk worden getijdencentrales beschreven en de wijzigingen en verbeteringen besproken.

*Prof. (Ret), J.Rudy Senten ScD, Higher Ind. Eng. Inst., Antwerp*
*Director (Ret), Sci. Lab. City of Antwerp (Belgium)*

German

## ZUSAMMENFAßUNG

Ozeangezeiten wurden durch die Attrakzion von Sonne und Mond auf die Erde verursacht. Es giebt Sondergezeiten, aber meistens hat mann nur zwei Gezeiten pro Tag. In bestimmten Regionen giebtst nur einmahl Hochwasser (z.B. Korea). Eine große Menge Energie wird beim Ebbe und Fluß freigelassen und könnte, möglicherweise umgesetzt in Kraft.

Die Fransözen haben daß einzige große Gezeitenkraftwerke in der Welt, in Britannien, auf der Rance Fluß, in der Nähe St. Malo's und Dinard gebaut. Hier wurden 240,000 kW Elektricität produziert seit 1968. Zur Zeit glaubte mann da.ß Gezeitskraftwerke unweltfreundlich wären, aber Heute hört mann Stimmen die hierüber nicht mehr einverstanden sein. Jetzt bleibt daß System doch nicht so schlimm daß Benutzung von die Gezeitenkraft keine Rolle spielen darft als Beischlag Elektrizität Brunne der Zukunft.

Vorlaufer der heutigen Gezeitenkraftwerke waren die Seeß-Muhle die dauernd Jahrhunderte mechanische Kraft dem Mensch besorgte. Wahrscheinlich wurden din erste See-Muhle auf Tigris und Euphrates Estuarium beim Bassora (Basra) gebaut. Weiter wird erzählt daß der Römischer-Byzantiner Feldherrn Belisarius, wann er in Rom belagen war, im V. Jahrhundert solche Muhlen benutzte.

In West-Europa finder man, folgens dem "Domesday Book" eine See- oder Gezeiten-Muhle am Eingang des Dover (Engeland) Hafens. Nachdem wurde in

Wales, Engeland, Frankreich, und auf Iberischen Halbinsel hunderten solche Muhlen gebaut. Europäische Einwanderer brachten die Technik nach Nordamerika, z.b.Neu-York.

Die Basistechnik vom See-Muhlen und Gezeitenkraftwerk its gleichartig. Leider ist es nur möglich eine kleine Menge Gezeitenenergie zu umformen in Elektrizität, selbst seit neue Turbinen daß benutzen von kleinere Ebbe-und-Fluß Unterschied ermöglicht hat. Weiter ist man heute auch mehr aufmerksam auf die horizontale Gezeitenbewegung, die man auch damals benützte für See-Muhle und die angewendet kan wurden bei Gezeitenkrafstellen. Solche Werke brauchen keinem Damm, und Damme sind daß teuerste Teil einer Zentrale.

Die kleine Kraftwerke im Kreisen Bristol (Engeland), Husum (Norddeutschland) und Suriname (Südamerika) sind seit manche Jahren verschwunden, und außer die Rance bestehen heute Pilot-Kraftwerke in Kanada (Neuschottland) und Rußland (Kislaya Bai), und scheinbahr annährend einhundert mini-Zentralen in Volksrepublik China. Ein größeres Kraftwerk wurde in Sdie Republik Korea abbestellt für politische Reden; aber nach 2005 Nahrichten sollen die Koreanen selbst die größte Gezeitenkrafwerke allerzeits beim Sihwasee bauen.

Über Werke in Australien, Argentinien, dem Vereinigten Staten ist noch wenig zu hören. Oder waren in die Regionen keine oder zu wenig Benutzer, oder politische Wiederstand endete alle Pläne. Schon wird in Internazionale Kongresse wieder über Ozeanenergie geredet und hört man von Japan, Mexiko, Brazil usw. Aber meistens sind Projekte von und kommen Voschläge auß. Engeland, Schottland und selbst Ierland. Ein Forschungszentrum wird aufgerichtet in die Orkney Inseln (Schottland) mit finanzielle Hilfe von Europa. Scheinbar sind die Ergebnisse gut und praktische Versuche wurden schon vereinbart.

Die Rance und Kislaya Werke sind 40-jahre alt. Sie haben "knolle Turbinen" eingebaut. Die ermöglichen Elektricitätproduktion beim Fluß und Ebbe. In Annapolis-Royal (Kanada) sind aber Rim-Straflo Turbinen eingestellt. Manche neue Technologien wurden seitdem im Bau- und Betrieb eingeführt. Durch daß unaufhörlich Öl- und Gas- preißsteigen und die nuklear Unwelt-sorge, können heute die Ozeanenergiebrunnen Kraft abliefern an einem Konkurrenzpreiß: Wind ist schon in, oder in der Nähe von, viele Meer-Orten benutzt. Wellen und Gezeiten wurden für Elektrizitätbesorgung in der Zukunft durch manche fuhrende Wissenschaftler stark empfehlt. Kommt Poseidon wirklich zur Hilfe?

Vielleicht ist die Dämmerung von Gezeitenkraftwerken dicht bei und sind die tatsächlich auß dem Schatten wieder nach vorn gekommen.

*V.D.*
*EFCA [European Federation of Consulting Engineers Assocaitions]*

## Polish

**RESUME**

*Co to są pływy?*

Pływy spowodowane są występowaniem sił grawitacyjnych pomiędzy ciałami niebieskimi i Ziemią. Ciała niebieskie, które w największym stopniu powodują pływy to Słońce i Księżyc, ze wskazaniem na Księżyc, który jako najbliższy Ziemi wywiera największy wpływ na zjawisko pływów.

Problemy w wytwarzaniu mocy z pływów są skutkiem występowania różnic w długości dnia słonecznego i księżycowego. Usiłowano zrekompensować straty wynikające z wymienionej różnicy, co zaowocowało pojawieniem się układów wykorzystywania pewnych form przechowywanej energii, tak by zachować pozorną ciągłość płynności wytwarzanej energii. W układach wykorzystuje się przechowywaną wodę, wodór albo sprężone powietrze.

Występują dwa rodzaje pływów: wzrost czyli przypływ i opadanie czyli odpływ; zalewanie wodą przypływu naniesionego prądami morskimi i odpływ mas wody.

Zjawisko pływów zostało opisane w dziesiątkach książek i artykułów. Stanowiło zagadkę przez wieki. Wyjaśnienie ich natury przypisywane jest Izaakowi Newtonowi.

Dynamika pływów była dyskutowana i wyjaśniana w powiązaniu z występowaniem powodzi w niedawno wznowionym wydaniu woluminów Wood'a.

W niniejszym rozdziale została przedstawiona krótka charakterystyka zjawiska pływów oraz przyczyn ich występowania. Dokonano także zwięzłego przeglądu różnych typów pływów.

W odniesieniu do możliwości działania elektrowni pływowych szczególne zainteresowanie skierowano na pływy półdobowe i dobowe.

Szacuje się, że moc uzyskiwana z pływów może być generowana za pośrednictwem różnego rodzaju elektrowni pływowych.

Istnieją elektrownie jedno- i dwuukładowe (pojedynczego lub wielokrotnego wykorzystania tej samej objętości (pojemności wody)) oraz jedno- i wielobasenowe.

W rozdziale pokazano kalkulację istotnych współczynników. Ponadto zostały streszczone podstawowe statystyki i podstawowe ich parametry.

*Prąd z prądów pływowych*

Człowiek przez wieki wykorzystywał energię uzyskiwaną w celu otrzymania mocy Prawdopodobnie posiał tą umiejętność już kilkaset lat przed naszą erą, ale w szczególności doskonalił ją od XI i w XII wieku, kiedy to wiatraki wykorzystujące energię pływów przyczyniły się do rozwoju przemysłu. Ówczesne wiatraki wykorzystywały zarówno horyzontalne pływy, jak i wertykalne.

Na przestrzeni lat, w swoich wysiłkach, człowiek wykorzystywał pływy do generowania energii elektrycznej w XX i XXI wieku skupiając się na elektrowniach pływowych. Elektrownie te wykorzystują wertykalne ruchy w pływach; rozwój dotyczył konstrukcji tam i wykorzystania zatrzymywania dorzecza albo wykorzystania obszarów retencyjnych.

Tama, wał albo zapora były finansową przeszkodą w konstruowaniu większej ilości elektrowni rozmiarów Rance albo nawet elektrowni rozmiarów rosyjskich czy kanadyjskich jednostek.

Zamiast wykorzystywać przypływy i odpływy, nie wykluczone ze będzie się w przyszłości wykorzystywać prądy pływów.

Do produkcji energii może być wykorzystywany zarówno odpływ jak i przypływ. Nie są potrzebne żadne zapory.

Tak więc, programy obecnie proponowane są nowoczesnym transponowaniem pracy współczesnych turbin pływowych. Mogą być one zbudowane za ułamek kosztów stacji zaporowej, ale ilość uzyskiwanej dzięki nim mocy jest skromniejsza.

Pilotowe stacje były wypróbowywane na różnych szerokościach geograficznych, prawdopodobnie najlepiej znana jest East River (Nowy Jork) obecnie w budowie.

*Udoskonalenia i postęp*

Od czasów gdy mniej niż pół wieku zostały zbudowane stacje Rance, Kislaya i Fundy poczyniono znaczne postępy w konstrukcji, obsłudze i doborze lokalizacji elektrowni pływowych. Generalnie zrezygnowano z użycia dużych koferdamów (duże wydatki poza budowaniem zapor).

Turbiny kształtowe bańkowe, rewolucyjny wynalazek wymyślony w czasie elektrowni Ranca, został zastąpiony przez turbinę Strablo. Ponadto rozważa się wciąż możliwości zastosowania innych typów turbin, szuka się nowych konstrukcji.

Wytwarzanie mocy zarówno z odpływów jak i prądów zatapiających nie jest rozpatrywane w kontekście ekonomicznym (angażowanie dodatkowych kosztów). Operacje przepompowywania są w fazie ponownego szacowania.

Postęp w elektronice przyczynił się do zmodyfikownia operacji i pozwolił na redukcję personelu. Wprowadzono także nową ochronę antykorozyjną.

Spośród przedstawionych propozycji tworzenia elektrowni w różnych regionach świata, żadne nie są realnie rozpatrywane z wyjątkiem ogromnej elektrowni w Republice Korei.

Wykorzystanie cech geomorfologicznych obszarów, jak na przykład istnienie półwyspów, przyczyniło się do powstania projektów które mogą być wprowadzone w życie.

Został zauważony trend by budować małe elektrownie (nawet przedstawiając ponownie zalety turbin pływowych), ze szczególnym uwzględnieniem rejonów w odległych i słabo zaludnionych regionach gdzie występują trudności z doprowadzeniem elektryczności. Elektrownie pływowe były rozpatrywane jako mające najgorszy wpływ na środowisko, słyszy się jednak głosy, że występują wątpliwości w dokładność oszacowania tego wpływu. Częstotliwość prowadzenia studiów nad tematyką wzrosła, pomimo faktu, iż autorzy pozostali sceptycznie nastawieni do zagadnień.


*Rance, Kislaya i inne*
    Elektrowni pływowa Rance River w 2006 roku obchodzi 40. lecie istnienia. Elektrownia Kislaya Bay w przyszłym roku osiągnie taki sam wiek. Także Hog's Island, Bay of Fundy zbliżają się do 40. urodzin. Wszystkie spisują się dobrze. Szacuje się, że jest ponad setka i kilkanaście małych elektrowni w Chinach, w kraju z ogromnym potencjałem do tworzenia elektrowni pływowych.
    Rozdział zawiera opis elektrowni i omówienie ich modyfikacji oraz rozwoju na przestrzeni ostatnich czterech dekad.

*Agata Krystosik-Gromadzinska Eng.*
*Assistant Professor & Department Chair*
*Technical University of Szczecin, Poland*


## Rumanian

**REZUMATE**

*Scoase la lumină din dosarele adormite în fişet*
    Charles de Gaulle trebuie creditat cu construirea primei mari centrale mareomotrice (acţionate de forţa mareei). Aceasta s-a întâmplat în urmă cu 40 de ani. Modelarea şi experimentări au fost întreprinse la Grenoble, Franţa la SOGREAH, Societé Grenobloise d'Applications Hydrauliques, care fusese şi urma să fie implicată în diferite studii privind centralele mareomotrice.
    Utilizarea energiilor mareei nu a fost un concept nou, de fapt unul foarte vechi, întrucât mori mareice au utilizat asemenea energii pentru a produce putere mecanică de milenii. Fluviul Rance a fost locul ales pentru construirea CEM Rance, şi multe mori mareice, încă în funcţiune, au trebuit demolate pentru a face loc unei centrale electrice. Centralele au fost luate în considerare în diferite areale geografice. Marea Britanie a avut în vedere fluviile Mersey şi Severen, Argentina a făcut studii de fesabilitate pentru Golful San José, Australia a luat în considerare Kimberleys, Coreea a avut în vedere câteva locuri ca Garolim sau Inchon, SUA chiar a început construirea unei centrale în regiunea Passamaguoddy şi Canada a examinat în mod repetat Bay of Fundy.
    Însuşi Leonardo da Vinci a proiectat planuri pentru utilizarea mareei ca sursă de energie. O literatură foarte abundentă a apărut şi conferinţe internaţionale privind utilizarea energiilor oceanului sunt frecvente în prezent. Suntem departe de importanţa minoră atribuită odinioară de către un preşedinte de conferinţă binevoitor doar câtorva lucrări cu acest subiect. Energiile oceanului şi forţa mareei

în particular sunt frecvent la ordinea zilei. Numărul firmelor implicate în cercetări *ad hoc* creşte continuu, de asemenea încercările pilot.

Observarea şi înregistrarea mareei datează din Timpurile Clasice – Herodot şi Euripide revin în memorie, Newton a oferit o explicaţie.

Pe lângă diferenţe ale mareei în privinţa temperaturii şi salinităţii, valuri, forţa curenţilor oceanului – şi a curenţilor mareici, vânturi marine, biomasa marină, toate sunt de interes ca surse de energie regenerabilă şi considerate benigne din punct de vedere ambiental. Centrala de pe fluviul Rance a transpus principiile străvechi ale morii mareice în lumea contemporană. Ea are un baraj, un bazin de captare (retenţie) de apă, turbine bulb (cu rezervor) şi este capabilă de generarea energiei electrice prin maree atât de flux cât şi de reflux. O şosea făcând legătura între cele două maluri ale canalului a fost amplasată deasupra. Din punct de vedere social şi economic facilitatea a promovat regiunea aproape uitată a Bretaniei în secolele XX şi XXI.

*Poseidon ne vine în ajutor*

Nu prea mult s-a spus la „Summit"-ul Johannesburg – 2002 despre rolul pe care alternativele energetice oferite de ocean în zonele costiere şi de larg le-ar putea juca în oprirea încălzirii globale. Realismul trebuie să predomine în ameliorarea statisticilor că forţa oceanului, în prezent, nu poate înlocui sursele tradiţionale de energie: ele sunt complementare, dar după cum s-a evidenţiat, orice (mică) fărâmă ajută.

Utilizarea vânturilor marine s-a extins considerabil şi prin mari „parcuri", îndeosebi de-a lungul coastelor Suediei, Danemarcei, Ţărilor de Jos, de exemplu, „ferme" mai modeste fiind înfiinţate, mai recent în largul coastelor Belgiei şi Statelor Unite. Germania, Marea Britanie, Irlanda nu se află mult în urmă.

Valurile au fost întrebuinţate pentru generarea de energie. Norvegia şi Marea Britanie conduc în această tehnologie dar portughezii au realizat poate (2006) cel mai mare proiect, utilizând, este adevărat, tehnologie britanică. Furia furtunilor şi opoziţia oficialilor din turism au condus în cele din urmă la necesitatea de utilizare a energiei nepoluante, dioxidul de carbon răspândit şi criza încălzirii globale constituind argumente însemnate.

Criticile asupra centralelor mareomotrice ca necesitând investiţii mari (şi trecând cu vederea longevitatea unor asemenea centrale comparativ cu durata de viaţă cu mult mai scurtă a centralelor tradiţionale şi chiar nucleare) şi insatisfacţia unor ambientalişti care invocă impactul barajelor, au determinat reîntoarcerea în trecut când morile mareice utilizau mareea. Este astfel posibil, odată redescoperită, să captezi energia mareei fără diguri şi baraje.

Problema mareei şi energiei sale va fi examinată în capitolele următoare. Dar trebuie menţionat că în prezent în afară de marea centrală Rance, instalaţii pilot au funcţionat în mod satisfăcător în Rusia, Canada şi China. Staţii locale modeste pot ajuta progresul în regiuni îndepărtate şi în zone mai puţin dezvoltate. Oare va renaşte moara mareică din „cenuşa sa" ca o pasăre Phoenix a secolului XXI?

Cuvintele sunt bune, dar faptele, desigur, sunt mult mai bune. Dacă energiile oceanului pot fi non-inepuizabile, hidrocarburile – indiferent dacă de pe

Annex VI: Summaries

uscat sau din mare – cu certitudine sunt rezerve consumate cu mare viteză. Vânturile, valurile, mareea au fost abordate cu succes. Alte surse ale oceanului precum hidrogenul, diferențele termice și salinitatea au avut un succes diferit. Proiecte uriașe sunt mai puțin probabile și unele dintre acestea ar putea crea probleme gigantice.

Energiile oceanului sunt considerate îndeosebi drept contributori ai producerii de electricitate, aplicații în domenii conexe, ca desalinizarea, geamanduri, pompe.

Din numărul limitat de operații de extracție relativ benigne din punct de vedere ambiental care sunt în mod simultan durabile, utilizarea energiilor oferite de către ocean a fost menționată adesea, în general cu rea credință, rareori întreprinsă. Nu este probabil irațional a reconsidera aceste surse de energie care, în lipsa unor proiecte grandioase, pot ușura povara grea proprie regiunilor îndepărtate, a țărilor sărace, a zonelor lipsite de hidrocarburi.
Inginerie medievală care a dăinuit

Morile mareice au jucat un rol important în dezvoltarea industrială, îndeosebi în Occident. Țărmuri presărate cu mori pe ambele laturi ale Oceanului Atlantic, unele amplasate chiar pe cele ale Pacificului. Deși înlocuite progresiv de tehnologii mai avansate, câteva au rămas în funcțiune și după cel de-al Doilea Război Mondial. *De facto* ele sunt precursorii centralelor mareice actuale. Moara mareică din timpurile de odinioară a inspirat visele inginerilor. Dacă numeroase locuri sunt adecvate pentru construcția de centrale, numărul lor a crescut în mod considerabil când turbine cu presiune joasă au apărut pe piață.

Lucrări despre moara mareică nu au lipsit dar cărțile cele mai remarcabile au apărut în ultima parte a secolului aparținând unor astfel de *aficionados* ca de la Vernhe, Boithias, Homualk de Lille. Nu mult mai târziu interesul pentru arheologia industrială, manifestat fie prin metoda amprentării, a reconstruirii dintr-o stare deteriorată sau nouă a dat naștere unor grupuri care și-au propus descoperirea, descrierea, restaurarea, chiar reutilizarea vechilor mori. Dacă asemenea mori pot fi găsite menționate pentru prima oară în Irakul timpurilor clasice, și unele au fost utilizate de către Belisarius când a fost asediat pe Tibru la Roma, mulți istorici se referă la menționarea morilor mareice în *Domesday Book* (Cartea Judecății de Apoi). Ele au devenit o componentă destul de comună a peisajului secolelor al 12-lea (Anglia, Țara Galilor) și al 13-lea (Țările de Jos). Multe au fost construite de-a lungul coastelor, dar unele au fost situate pe fluvii (de ex. Dunărea), sau chiar sub poduri (Tamisa), în timp ce altele au făcut uz de peninsule sau au fost înălțate pe insule. Francezii au introdus tehnica în Canada, olandezii în regiunea New York, iar alții în Massachusetts și Maine.

Morile au avut uneori un mic lac de acumulare, alteori numai o roată activată de curentul mareic al fluviului. Mecanisme destul de sofisticate au fost găsite în interiorul morilor. Energia era transformată îndeosebi în putere mecanică, deși pe lângă măcinare, unele mori au fost utilizate pentru producția de bere, gatere, dispozitive de ridicare (macarale).

Mori care prezintă și rețin interesul în prezent sunt cele din Marea Britanie, Franța, Portugalia, Spania, America de Nord – unde odinioară peste 300 dintre ele au funcționat – și, într-o măsură mai mică, cele din Țările de Jos.

Au existat comentarii asupra „redeșteptării" sau „învierii" morilor mareice și a utilității lor posibile în producerea de energie pentru zone îndepărtate și izolate.

Ce este mareea?

Mareea se datorează atracției exercitate de către corpurile cerești asupra Pământului. Principalele corpuri având acest efect sunt Soarele și Luna, aceasta din urmă fiind cea mai apropiată de pământ și având impactul major. Diferența între ziua lunară și solară ridică anumite probleme generării de energie mareică. Eforturile de compensare a acestora se află la originea schemelor de resincronizare. Acestea pot include depozitarea apei, utilizarea hidrogenului sau aerului comprimat.

Mișcarea mareică este dublă: ridicarea și coborârea nivelului apei, și deplasarea orizontală a apei: fluxul și refluxul.

Fenomenul mareei a fost descris în extrem de numeroase cărți și lucrări. A încurcat savanții timp de milenii. Isac Newton este creditat cu explicarea sa. Dinamica mareei a fost discutată și explicată, printre altele în legătură cu inundațiile, *ad libitum* în volumele de referință republicate recent ale lui Wood. Acest capitol oferă o discuție sumară asupra cauzelor și caracteristicilor iar diferitele tipuri de maree sunt prezentate pe scurt. Mareele semi-diurne și diurne interesează în privința centralelor electrice mareice.

Generarea energiei mareice este explicată astfel încât diferitele tipuri de centrale mareomotrice să poată fi evaluate. Acestea includ sisteme utilizând una respectiv două direcții ale mareei și sisteme utilizând unul sau mai multe bazine de captare (retenție). Se prezintă calcularea coeficienților relevanți. Au fost rezumate statisticile de bază și parametri statistici implicați.

*Rance, Kislay, Fundy ș.a.*

Centrala mareomotrică Rance și-a sărbătorit cea de a 40-a aniversare în 2006. Centrala Kislay Bay este în curs de a atinge aceeași vârstă peste încă un an și Hog's Island, Bay of Fundy, are aproape aceeași vârstă. Toate s-au comportat bine, chiar dacă nici una nu a declanșat o serie de construcții de centrale mari sau chiar modeste. Se presupune că există peste o sută de centrale mareomotrice mici, și câteva întrucâtva mai mari, în China, o țară cu un potențial energetic mareic uriaș.

Capitolul oferă o descriere a centralelor și o discuție asupra modificărilor și ameliorărilor care au avut loc în cursul ultimelor patru decenii.

*Curent electric din curenți mareici.*

Energia mareei a fost utilizată de către om să producă putere mecanică de milenii, poate cu câteva ere dinaintea celei prezentate, dar deabea din secolele al XI-lea și al XII-lea morile mareice au contribuit la progresul industrial. Acele mori au făcut uz uneori de mișcarea orizontală a mareei, alteori de cea verticală.

În eforturile sale de valorificare a mareelor pentru generarea de electricitate, omul s-a concentrat în secolele al XX-lea și al XXI-lea asupra centralelor mareomotrice care valorifică mișcarea verticală și care au implicat construirea unui dig și utilizarea unui bazin de captare (retenție) sau a unei zone închise. Digul, stăvilarul sau barajul au constituit un impediment financiar în construirea unor centrale mareomotrice de tipul celei de pe fluviul Rance, sau chiar a celor ruseşti sau canadiene în acest sens. În loc de a controla creşterea şi scăderea mareei, este preferabilă utilizarea curentului mareic. Ambii curenţi ai fluxului şi refluxului pot fi utilizaţi. Barajul nu este necesar.

Astfel, diferite scheme propuse în mod curent sunt o transpunere modernă a funcţionării morii mareice obişnuite. Acestea pot fi construite cu o fracţie din costul unei staţii cu baraj, dar cantitatea de energie obţinută este mai modestă. Staţii pilot au fost încercate în diferite zone geografice şi poate cea mai cunoscută este centrala de pe East River (New York City) aflată în construcţie.

*Alexandru S.Bologa PhD*
*Scientific Director,*
*Romanian National Institute for Marine Research „Grigore Antipa"*
*[former] Director, Black Sea Station International Ocean Institute*

*Russian*

Резюме

*Введение – Из чулана – снова на сцену*
Честь создания первой современной большой приливной энергетической установки принадлежит Шарлю Де Голлю. Это случилось всего каких-нибудь 40 лет назад. Моделирование и первые испытания проводились в Гренобле (Франция) на базе SOGREAH – общества прикладной гидравлики Гренобля, участвовавшего в целом ряде исследований по приливным электростанциям.

Использование энергии приливов не было новым понятием, а фактически - даже очень старым, поскольку водяные мельницы использовали подобную энергии для обеспечения механической работы уже в течение тысячелетий. Река Ранс была местом, выбранным для возведения приливной электростанции, и много водяных мельниц, всё еще работавших, должны были быть демонтированы, чтобы освободить для неё место. Рассматривались и другие географические местоположения. Великобритания присматривалась к рекам Мерси и Северн, Аргентина провела технико-экономическое обоснование для залива Сан-Хосе, Австралия рассматривала Кимберли, Корея имела на примете несколько мест, таких как Гаролим или Инчхона, США даже начали строить станцию в области Пассамакуодди, а Канада неоднократно исследовала залив Фанди.

Даже Леонардо да Винчи разрабатывал планы использовать приливов как источника энергии. Много работ опубликовано, и в настоящее время много международных конференций посвящено использованию энергии океана. Мы уже далеки от ситуации, как когда-то однажды доброжелательный председатель конференции дал возможность представить несколько статей по этому предмету. Океанская энергия, и приливно-отливная энергия в частности – теперь часто находятся на повестке дня. Устойчиво увеличивается число фирм, занятых специализированными исследованиями и экспериментальными разработками.

Наблюдение и регистрация приливов идут с классических времён - приходят на ум имена Геродота и Еврипида, а Ньютон дал им объяснение.

Помимо приливов, различия в температуре и солености, волны, сила океанских течений (и приливно-отливных течений, и приливных струй у Великобритании), и морские ветры, и морская биомасса - все рассматривается как возобновляемые источники энергии, причём экологически щадящие. Речная электростанция на Рансе перенесла старые принципы водяных мельниц в современный мир. Она имеет плотину, сдерживающий бассейн, капсульную гидротурбину и способна производить электроэнергию и на отливе, и на приливе. Шоссе, связывающее обе стороны канала, помещено сверху на плотине. Социально и экономически, станция подняла в XX и XXI столетия довольно заброшенную область Британии.

Глава Посейдон приходит на помощь
На Встрече на высшем уровне в Иоханнесберге в 2002 году было сказано не много о роли, которую могут сыграть альтернативные источники энергии, предлагаемые океаном на берегу и в прибрежных водах, в сдерживании глобального потепления климата. Реализм должен преобладать в оценке ошеломляющей статистики, утверждающей, что энергия океана в настоящее время не может заменить традиционные источники энергии: они дополняют друг друга, но, как было указано, даже небольшой вклад - уже помогает.

Значительно расширились использование морских ветров, и помимо огромных "ветропарков", особенно на побережьях Швеции, Дании, Нидерландов, устанавливаются, например, и более скромные "фермы" – в море около Бельгии и Соединенных Штатов. Германия, Великобритания, Ирландия отстают лишь незначительно.

Используется и энергия волн. Норвегия и Великобритания - лидеры в этой области, но честь создания (2006) наибольшей системы принадлежит португальцам, хотя и использовавшим британскую технологию. Ярость штормов и оппозиция чиновников от туризма наконец уступили потребности использования энергии, не загрязняющей окружающую среду, а современные концентрации углекислого газа и кризис глобального потепления явились сильными аргументами.

Неудовлетворённость большой капиталоёмкостью приливно-отливных электростанций (невзирая на их долговечность по сравнению с намного более коротко-живущими традиционными и даже ядерными станциями), и несогласие регионов с воздействием требующихся плотин стали поддержкой возвращения к прошлому, когда приливно-отливные течения использовались приливными мельницами. Таким образом, мы открыли заново, что приливно-отливную энергию можно захватить и без плотин или заграждений.

Вопрос приливов и их энергии будет исследован в последующих главах. Следует сказать, что в настоящее время, помимо большой станции на Рансе, экспериментальные сооружения удовлетворительно показали себя в России, Канаде и Китае. Скромные местные станции смогли помочь развитию отдаленных и слабых областей. Возродятся ли приливные мельницы «из пепла», как птица Феникс XXI столетия?

Слова хороши, но действия, конечно, намного лучше. Если энергия океана может быть неисчерпаемой, то углеводороды – с суши ли, или из моря – определённо будут исчерпаны уже вскоре. Ветер, волны, приливы уже успешно используются. Другие океанские источники – такие как водород, разница в температурах и солёностях – встречены с неравным успехом. Огромные системы менее вероятны, и среди них некоторые могут породить гигантские проблемы.

Если энергия океана видится особенно значимой в производстве электроэнергии, то должны развиваться и приложения в соответствующих областях, таких как опреснение, буи, насосы.

Из довольно ограниченного числа постоянно существующих и относительно щадящих по воздействию на окружающую среду источников, собирание энергии, которую даёт океан, упоминается часто, часто критикуется, но редко предпринимается. Но вероятно, это не столь уж нелепо - посмотреть снова на эти источники энергии, которые – если грандиозные планы не предполагаются – могут облегчить тяжёлое бремя отдалённых регионов, бедных стран, лишенных углеводородов областей.

Глава Средневековые разработки, которые продолжаются
Приливные мельницы играли важную роль в индустриальном развитии, особенно на Западе. Мельницами испещрены побережья с обеих сторон Атлантического океана, некоторые расположены даже на Тихом океане. Хотя они постепенно вытеснялись более передовыми технологиями,

несколько мельниц продолжали работать много лет спустя после окончания Второй мировой войны. Фактически они - предшественники сегодняшних приливно-отливных электростанций. Приливная мельница прошлых времен вдохновила мечты инженеров. Многие участки являются подходящими для установления станций, и их число значительно выросло, когда на рынок были выпущены низко-напорные.

Нет недостатка в статьях по приливным мельницам, но наиболее замечательные книги вышли в последней части столетия за авторством таких приверженцев как Верн, Бойтиас, Хомалк де Лилль. Не намного позже интерес к индустриальной археологии – в миниатюре ли, восстановленного, оставленного или созданного заново оборудования - обусловил образование групп, взявшихся за исследование, описание, восстановление, и даже использование заново старых мельниц. Хотя подобные мельницы могут быть прослежены в прошлое до классических времен Ирака, а некоторые были построены ещё Белизариусом, осаждённым на реке Тибр в Риме, большинство историков ссылаются на упоминание о морских мельницах в Книге Судного Дня. Они стали весьма общей частью пейзажа в 12-ом (Англия, Уэльс) и 13-ом (Нидерланды) столетиях. Большинство строилось на побережьях, но некоторые были расположены на реках (например. Дунай), или даже под мостами (Темза), другие использовали полуострова или были возведены на островах. Французы принесли технику к Канаде, голландцы - в область Нью-Йорка, другие к штату Массачусетс и Мэн.

Мельницы имели иногда сдерживающий водоем, иногда только колесо, крутящееся потоком реки. Внутри мельниц были найдены довольно сложные механизмы. Энергия главным образом преобразовывалась в механическую работу, хотя помимо размола, некоторые мельницы использовались на пивоварнях, лесопилках, подъёмниках, и т.д.

Мельницы, которые представляют интерес для исследователя до настоящего времени - это те что в Англии, Франции, Португалии, Испании, Северной Америке, где однажды более 300 из них работали – и, в меньшей степени, те что в Дании и Голландии.

Имеются комментарии о «возрождении» и «восстановлении» приливных мельниц и их возможной пользе для производства энергии для отдалённых и изолированных мест.

*Irina Chubarenko PhD*
*P.P. Shirshov Institute of Oceanography*
*Russian Academy of Sciences*

## *Spanish*

## RESUMEN

*¿Qué es una marea?*
El movimiento de la marea se origina debido a la atracción ejercida por cuerpos celestiales sobre la tierra. Los principales cuerpos que tienen este efecto son el sol y la luna; esta última que es la más cercana a la tierra, es la que tiene el mayor impacto. La diferencia entre el día lunar y solar ha representado algunos problemas a la generación del poder de la marea. Los esfuerzos para compensar dicho problema deben involucrar almacenamiento de agua, uno de hidrógeno y aire comprimido.

El movimiento de la marea es doble: el levantamiento y la caída del nivel del agua, y el desplazamiento horizontal del agua: la corriente que sube o inundación y la corriente que baja.

El fenómeno de la marea ha sido descrito en diversos libros y artículos. Ha sido un enigma para científicos durante milenios. Isaac Newton ha sido uno de los tantos que han explicado dicho fenómeno. Éste ha sido discutido y explicado en conexión con inundaciones, ad limitum, en las recientes re-publicaciones de volúmenes de referencia por Wood. Los diferentes tipos de mareas brevemente revisadas y una discusión corta de sus causas y características es aportada en este capítulo.

La generación del poder de la marea es explicada de tal manera que los diversos tipos de poder de mareas pueda ser valorada. Estos incluyen un efecto sencillo y doble en esquemas. Calculaciones de coeficientes relevantes es mostrado. Las estadísticas básicas y parámetros básicos envueltos han sido sumariados.

*La corriente desde la corriente de la marea* 4.
La energía ha sido utilizada por el hombre para proveer poder mecánico durante milenios, tal vez mucho antes que la era de la corriente; pero es particularmente desde el siglo XI y XII cuando las máquinas de marea contribuyeron al progreso industrial. Dichas máquinas hicieron uso del movimiento horizontal y vertical.

En el siglo XX y XXI, en los esfuerzos para que la marea genere electricidad, el hombre se ha enfocado en tpp-s que usó el movimiento vertical y que involucró la construcción de una presa. La presa, dique o barrera, ha sido un impedimento financiero a la construcción de más tpp-s del tamaño del Rance, o incluso hasta de los rusos o canadienses para dicha materia. En lugar de intervenir con ligeros golpes el levantamiento y caída de mareas, es concebible usar la corriente marea. Ambas corrientes pueden ser usadas. La barrera no es necesaria.

De esta manera, diversos esquemas actualmente propuestos se encuentran en una moderna transposición de carrera de generadores de mareas. Estas pueden ser construidas en una fracción del costo de una barrera, pero la cantidad de poder obtenido es más modesto. Las estaciones pilotos han sido tratadas en varias ubicaciones geográficas y tal vez el mejor conocido es el Río del Este en Nueva Cork; actualmente bajo construcción.

*Mejoras y progresos*
Las grandes zancadas han sido hechas en construcción, operación y emplazamiento del poder de la marea desde las estaciones de Rance, Kislaya y Fundyn que han sido construidas hace menos de la mitad del siglo. El uso de "cofferdams" (batardeaux), otro gasto además de la construcción de barreras, ha sido generalmente pospuesto. La turbina bulbo, un invento revolucionario en la época del tpp Rance fue concebida, ha sido suplida en algunos esquemas por la turbina Starflo y otros estilos que aún están en consideración.

La generación de poder para corrientes que bajan y aumentan, no está vista tan económicamente justificada con los extra costos involucrados, y operaciones de bombeo están bajo re-evaluaciones. El progreso en electrónicos han modificado operaciones y han permitido la reducción de personal. Nuevas protecciones anti corrosivas han sido introducidas.

Si las propuestas han sido colocadas sobre la mesa que está observando enormes plantas en varias regiones del mundo, ninguna es seriamente considerada, excepto para una planta mayor en la República de Corea. El uso de particularidades geomorfológicos, tales como penínsulas, han generado proyedctos que pueden liderar la implementación. Una cierta tendencia a construir pequeñas plantas, incluso reintroduciendo el generador de marea, ha sido enunciada con un ojo de poder de producción en regiones remotamente pobladas donde la electricidad central sería difícil de situar.

Mientras que el ttp ha sido considerado, en el peor de los casos, como algo que tiene un impacto benigno en el ambiente, lo que ha puesto en duda la precisión de los estudios de evaluación en la materia.

*Rance, Kislaya, Fundy*
La planta de poder de marea del río Rance celebró su cuarenta aniversario en el 2006. La planta de la bahía de Kislaya está a punto de alcanzar la misma edad en un año más; al igual que la isla Hog y la bahía de Fundy.

Todas han realizado un buen trabajo, incluso si ninguno ha inflamado el torbellino de construcción de plantas grandes. Se supone que hay alrededor de cientos de pequeñas plantas y unas cuantas grandes tpp-s en China, un país con un enorme potencial de poder de mareas.

El capítulo provee una descripción de las plantas y una discusión de las modificaciones y mejoras hechas alrededor de las últimas cuatro décadas.

# Annex VI: Summaries

## *Italian*

### Resumo

Le variazoni del livello del mare possono essere sfruttate per creare energia potenziale, cioè l'energia che consente a un corpo—l'acqua—di compiere un lavoro in virtù della sua posizione. Esistono numerosi siti adatti alla costruzione di centrali maremotrici. Sebbene I costi primary siano elevati, gli economisti ritengono che il costo dell'energia non inquiante di marea si stia rapidamente avvicinando a quello delle centrali termiche tradizionali che inquinano l'ambiente.

Le maree si alzano e si abassano principalmente per effetto dell'attrazione lunare e solare. Un grosso svantaggio nella generazione dell'energia di marea è dato dal fatto che le maree sono legate al ciclo lunare che esse variano nel corso dell'anno. È possibile eliminare in parte questo svantaggio con artifici tecnici e tarare l'energia potenziale acummulata quando non si utilizza la potenza di mare.

Le prime installazioni "moderne" per la sfruttamento della forza di marea risalgono all secolo mille. Sono gli mulini a marea, solitamenti installati in corrispondenza di un estuario dove l'acqua veniva trattenuta con l'alta marea e poi rilasciata al verificarsi della bassa marea attraverso uno stratto canale, qui il mulino faceva girare una ruota idraulica. Sono numerosi I siti che si prestano all'installazione di centrali maremotrici.

Una stazione mareomotrici richiede uno sbarramento o diga in un'area in cui l'ampiezza di marea è alta. La maggiore centrale mareomotrice oggi esistente è sul fiume Rance in Bretagna. Le turbine sono dell typo bulbo che permesse il funzionamento bidirezionale, con l'impiego di un systema di pompa e turbine.

Il costo dei cassoni (inglese cofferdams, francese batardeaux) fu pari a un terzo della spesa prevista per la costruzione della centrale della Rance. Per la sua piccola centrale (Baia di Kislaya, presso Murmansk) I sovietici hanno eliminato tale costo e la fase di erezione mediante l'uso di strutture prefabricate di calcestruzzo, trasportate via acqua in situ e autoaffondate su fondazione pre-allestita.

L'energia di marea è un sistema di produzione dell'elettricità que, secondo le opinione prevalenti, è già maturo per l'attuazione. Il maggiore ostacolo è dato dai costi. I crescenti prezzi del petrolio, gas, li probleme di emissione de $CO_2$, il cambio climatologico, la maggior durata di una centrale mareomotrici, il progresso tecnologico nella trasmissione dell'energia, e la simplicazione della costruzione sono tutti fattori che controbuiscono a considerare l'energia mareomotrici come fonte energetica alternativa or sussidiaria.

# الطاقه من المد والجزر
## تأليف شارلييه و فينكل

يسرنى أن أقدم للقراء فى العالم العربى كتاب الزميلين شارلييه و فينكل عن ظاهرة المد والجزر فى البحار , واستغلال هذه الحركه الدائمه فى انتاج الطاقه.

ويرجع استعمال طواحين المد والجزر الى العصور القديمه. ويذكر العراق كمكان ظهرت فيه النماذج الأولى لهذه الطواحين التى استخدمت فى انتاج القوه الميكانيكيه بدلا من الأعتماد على قوى الحيوان أو الأنسان. وقد أنتقلت هذه الوسائل التكنولوجيه التقليديه لاستغلال طاقة البحر فى انتاج القوه الميكانيكيه الى انجلترا و ويلز وسواحل فرنسا واسبانيا والبرتغال المطله على المحيط الأطلنتى. وأخذ المهاجرون الأوربيون هذه الوسائل التكنولوجيه معهم الى شمال امريكا حيث أقاموا طواحين المد والجزر على سواحل كندا وعلى طول سواحل المحيط الأطلنتى جنوبا حتى ولاية جورجيا الامريكيه.

وتعتبر هذه الطواحين النماذج الأوليه التى سبقت محطات المد والجزر الحديثه لأنتاج القوه الكهربائيه. ومع ارتفاع اسعار البترول يتزايد الحديث بطريقه جديه للعوده الى هذه النماذج التقليديه , ولكن هذه المره لأنتاج الكهرباء بدلا من القوه الميكانيكيه. وتتناسب هذه المحطات الصغيره مع احتياجات المناطق النائيه أو القليلة السكان , حيث يكون انشاء محطات القوى الكهربائيه من المد والجزر غير مجديه من الناحيه الأقتصاديه. وقد بنيت محطات صغيرة الحجم فى بعض المواقع. وتحتل الصين المركز الاول من حيث عدد هذه المنشآت الصغيرة الحجم.

وظاهرة المد والجزر تتمثل فى نوعين من الحركه : حركه رأسيه ارتفاعا وانخفاضا وحركه أفقيه لتيار منعكس من البحر الى اليابسه ومن اليابسه الى البحر ويتسبب فرق التوقيت بين اليوم الشمسى واليوم القمرى فى خلق مشكلات فى انتاج الكهرباء من المد والجزر. وقد أمكن التغلب على هذه المشكلات بتطبيق بعض الحلول التكنولوجيه مثل ضخ المياه وتخزينها لأستعمالها فى أوقات الحاجه.

ويلعب المد والجزر النصف يومى والمد والجزر اليومى دورا كبيرا فى عمليات توليد القوه الكهربائيه من المد والجزر. ويحتوى الفصل الخاص بتوليد الكهرباء على وصف لمحطات القوى الكهربائية وبشرح الانتاج الكهربائى ويقارن بين الانواع المختلفه لمحطات القوى الكهربائيه من المد والجزر ويوفر المعلومات الحسابيه اللازمه لمتابعة الموضوع.

وقد تركزت خطط بناء محطات القوى الكهربائيه من المد والجزر خلال القرنين العشرين والحادى والعشرين على مشروعات كبيره تقتضى بناء سدود وأحواض لتخزين المياه , وهذه السدود والخزانات ترفع من التكلفه الفعليه وتثقل كاهل هذه المشروعات. لذلك اتجهت مشروعات مستحدثه الى محطات لا تستلزم انشاء سدود باهظة التكاليف. وبالطبع فان الاقتصاد فى التكاليف صاحبه انخفاض فى انتاج القوى الكهربائيه. ويزداد الأهتمام حاليا بتيارات المد والجزر كمصدر متواضع للقوى الكهربائيه.

*Irasema Guzman*
*Dipl. Ecole Sci. Pol. & Sociales, Université de Lille, France*
*Cand. Dr Jur., Fac. Jur. Univers. Autonoma de Mexico*

# Mini-Glossary*

**Air-storage, compressed**: air is compressed and stored at pressure in underground caverns during off-peak hours of a tidal power-plant (tpp).
**Amplitude [tidal]:** the half of the constituent tide; occasionally used synonymously to range (France) (see also **range**).
**Annual capacity factor**: quotient of average output and installed capacity (usually expressed in MW).
**Apogean tide:** tide of decreased range occurring each month near the time of the moon's apogee (antonym of **perigean tide** that has an increased range).
**Barrage:** dam; here its function is to control the flow of water, not to stop it totally.
**Benefit/cost ratio:** comparison of the economic benefits of a project with its costs.
**Biofouling:** fouling caused by growth or marine organisms
**Bore**: a high breaking wave of water, advancing rapidly up an estuary.
**Bulb turbine:** olive-shaped axial turbine that can function both as turbine and pump.
**Caisson:** water-tight box used as a gate; *also* box wherein people can work under water; *or* a caisson designed to be placed at each end of the main structure to retain the rock-fill dam. In a tidal power station a large reinforced pre-stressed concrete structure that can be constructed on dry land, usually near the coast, then launched, floated and towed into position.
**Capacity factor:** ratio of the amount of product (electrical energy) actually produced in a unit of time to its maximum production rate over that period [synonymous to **local capacity**]
**Change of tide**: reversal of the direction of motion of a tide
**Cofferdam:** watertight temporary structure built on a river or lake bottom to allow working on the dry during laying of dam or barrage foundation
**Cribs**: caissons (q.v.) or structures not containing powerhouse[s] nor sluice gate[s]
**Darrieus turbine**: vertical-axis turbine, invented 1925, consisting of blades with airfoil cross section.

---

*Some words included in the glossary have several meanings; only the meaning relevant to tidal matters has been retained.

**Dead dike** (dyke): dike containing neither powerhouse nor machinery
**Dependable peaking capacity**: amount of power that can be produced, upon demand continuously during no less than 4 hours, to meet daily peak load requirements.
**Dike (dyke):** embankment that may control or confine water
**Diurnal**: daily; **diurnal inequality**; difference in heights and duration of two successive high or low waters, or, of speed and direction of two flood or ebb currents of each day; **diurnal range**: amount of variation between maximum and minimum water level during 24 hours; **diurnal tide**: tide with only one high and one low water each lunar day.**Double-effect:** see reversing operation
**Economic rent**: income beyond that needed to bring or keep a firm in operation.
**Electric grid**: general electric transmission system of a region
**Flow**: combination of tidal and non-tidal currents.
**Head**: distance water can be dropped to recover its potential energy.
**Installed capacity**: total nameplate-rating of turbogenerators.
**Off-peak:** period of low demand (of electric power)
**Periodicity:** ranges, for tides, from 3.1 hours to 1600 years.
**Powerhouse:** structure containing a turbogenerator and associated equipment.
**Pumping process:** in tidal schemes, using pumps to over-fill or over-empty the retaining basin; pumped water storage consists in pumping water to an upper reservoir during off-peak hours and letting it run later through a turbine to a lower reservoir generating electricity at peak demand.
**Range**: difference of height of water level between high and low tide (see **amplitude**)
**Reversing operation** (synonym of **double-effect**): in a tidal power scheme electricity using both ebb and flood currents. In a **single-effect** system generation takes place only with water running in a single (usually seawards) direction.
**Rim turbine:** the rotor surrounds the turbine runner as a rim.
**Tidal regime:** mode of behavior of tidal system, including time and extent of rise and fall and associated tidal currents.
**Unretimed:** a generating tpp where no retiming takes places. **Retiming** uses some form of energy storage so that the output is smoothed into some semblance of continuity.
**Venturi tube:** short tube with constricted throat-like passage that increases velocity and decreases pressure on a fluid forced to pass through it.

# Index

Aber Wrac'h, France, failed tidal mill, Rance River power station, 105
Aero-generator, 5
AeroVironment, 146
Agger (double tide), 66
AGRA program, 113, 208
Al-Magdisi Shams al-Din, Arab geographer, 31
Alternative energy sources, 26
Ancillo mill, Santoñary, Cantabria, 61
Annapolis-Royal Pilot plant (Canada), 85, 121, 213, 214
Annual theoretical production (in kWh), 68
Anomalistic tides, 66
Apogean tidal current, 71
Appleyard, 137
Aqua power barge, 146
Archimedes Wave Swing generator, 12
'A rodete, rodicio, ' horizontal wheel, 43
Arillo tide mill, Cadiz, San Fernando–Cadiz road, 54, 55, 59
Arklow Bank, 6
Auto-correlation, serial correlation, 73
Auto-regression, statistical technique, 73
Auto-regressive series, 72
AWCG, tidal stream device, 150
Axial-flow propeller machines, 112

Barrage
  highways on top, 132
  plant, electricity production, 111, 142
Bartivas tide mill, Chicanadela Frontera, 59
Bay of Fundy
  estuaries, 156
  fuel cost savings, 158
  Tidal Power Review Board, 213
  tides, 2, 3

Bi-directional ducted horizontal turbine, 215
Birdham tide mill, Sussex, UK, 50
Birlot Mill, Isle of Brehat, France, renovated tide mill, 36
Bishopston mill, Great Britain, 41
Blue Energy, 210
Brooklyn Mill, New York, USA, 38
Broome, Western Australia, 81
Bulb turbine
  generating caisson, 98
  groups, Rance River plant, 83, 109
  high capacity, 116
  reversible, 115
  types, 107, 213

Carew, Pembrokeshire, 49, 50
Cathodic conversion, 137
*Centrales marémotrices*, suitable sites for tidal power generation, 31
Centre for Sustainable Water Technology, The Netherlands, 129
Channel tides, tapping of, 131
Chausey Island proposed tidal power scheme (France), 3, 108
Chelsea mill, Massachusetts, USA, 38
Chinese plants, 85
Climate change, 134, 211
$CO_2$-polluters, 139
Coastal protection, 46
Coastal zone alternative energy, 19
Cobscook Bay, Falls Island, 122
Cofferdam, 85
Co-generation, 105, 129
Compact tidal power stations, 127
Concerted Action on Offshore Wind Energy in Europe (CA-OWEE), 5
Converging wave channels, 9

257

Coriolis
  deflection, 68
  Project, 146
Current, from tidal current, 141

Daily lunar retardation, 66
Darrieus turbine, 126, 133, 147, 151, 214
Davis Hydro Turbine, 15
Demi-Ville (France), dual-powered mill, 146
Derrien Rock Bridge, France, Tide mill, 35
Dikes, 40
Diurnal tide, 66
Doomsday date, 134
Double-effect
  generation, tidal power, 214
  mill, use incoming and outgoing tides, 39
Dual-powered tidal current mills, 112
Dunkirk 'Perse mill', 146
Dynamic dams, 149

Earth day, 66
East Medina mill, Wippingham, Isle of Wright, 49
Ebb- and flood-currents, to generate electricity, 107
Ecosystem research, 156
Electrical power generation
  aero-generator, 5
  Archimedes Wave Swing generator, 12
  axial-flow propeller machines, 112
  bulb turbine, 115, 214
  Eleicité de France, 110, 113, 156, 219
  General Electric Company, 120
  hydro-electric power stations, 109
  Kinetic Hydro Energy Conversion Systems, 147
  Kinetic Power Hydro Systems, 218
  ocean alternative energy, 104
  OTEC, 1, 13, 22, 23, 27
  Salford Transverse Oscillator, 148
  tidal power, 80, 87, 108
  turbines, 15, 116, 150, 206, 209
Electricité de France, 151, 219
Eling mill (reconstructed operating tide mill), Southampton, UK, 56, 57
Eling Tide Mill Trust, Southampton, 46
EMEC (European Marine Energy Center), Orkney, Scotland, 218
Energy potential, 143
Entrance sluice gates, 40
Environmental
  impact assessments, 155
  implications, tidal mills, 154
EPRI (Electric Power Research Institute), 210

Equinoctial
  spring tide, 66
  tides, 66
European Marine Energy Center, 131
Ewing Mill, England, working museum, 36
Exit gates, 42
Extractable power, 144
Extra-low head tidal power, 133

First Argentine Congress of Ocean Energy, 123
Fish pathway, 83
Fishways, Annapolis-Royal, 156
Florida Current, potential energy, 146
Flow-of-the-river potential, 143
Fluid flow energy conversion, 126
Fourier analysis, 75
Fromveur passage, Brittany, exploitable power, 145

Gates to the sea, portes de la mer, France, 41
General Electric Company, 120
Gibrat-Lewis-Wickert equation, 69
Gorlov's barrier, 125
Grain stores, 44
Grand Traouieros mill, Côtes d'Armour, 58
Green tides, 110
Grinding stones, 44
Grist mills, 44
Gulf Stream, power generation, 132

Hammerfest (Norway), submarine station turbine, 151
Harmonic analysis, 71
Harnessing tides, 69
Harza's axial flow turbine, 214
Hawaii Ocean Science and Technology Park, Hawaii, USA, 14
Helic[oid]al turbine, 209
Herrera mill, Bay of Cadiz, 41
Hi-Spec Research, 137
Hog's Island, Bay of Fundy, 85
Horizontal axis turbine, 149, 208
Huaneng Dandong Power Plant, China, 136
HVDC bus, 207
Hydraulic wheels, overshot, midshot, undershot, 39
Hydro-electric power stations, classic type, 109
Hydrogen
  hydropower development, 138
  injected into salt deposits, 72
  power, 17
Hydroxide-methyl-furfural, 18

Index 259

*Kaimei*, barge with compressed air chambers, 10
Kaplan turbine, 109
Keroilio mill, Plougoumelen, Morbihan, 61
KHECS (Kinetic Hydro Energy Conversion Systems), 147
Kianghsia facility, China, 85
Kimberley Region tidal power plant project (Western Australia), 101
Kinetic Power Hydro System (KHPS), 218
Kisgalobskaia plant, environmental assessment, 154
Kislaya Bay tidal power plant (Russia), 84, 91, 92, 118
Kislogubsk plant, Russia, 110
Kyoto agreements, 127

La Rance tidal power plant
    alternative operating modes, 76
    modes of operation, 110
Larger lunar elliptic ($N_2$) tidal component, 71
Large Russian tidal power plants, potential sites for, 96
Leonardo da Vinci, notebooks of, 103
Leqing Bay, China, tidal power plant, 96
Linear
    potential, specific region, 146
    vernier hybrid permanent magnet machine, 136
Lloyd Energy Systems, 131
Lockheed OTEC scheme, 27
London Bridge, 44
LonWorks Fieldbus digital communication network, 135, 208
Low-head hydro-electric plant, Pointe Pescade, Sidi Ferruch, 10
Low-pressure air turbine technology, 158
Lunar
    day, 66
    fortnightly ($M_1$) tidal component, 71
Lunar-raising force, 66
Lunisolar
    diurnal tidal component ($K_1$), 71
    elliptic ($K_2$) tidal component, 71
Lynmouth turbine, 149

Marine
    biomass conversion, 14
    currents, 14, 15
    Current Turbines, 209, 218
Marine
    wind farm, 5
    winds, 4
Mean discharge, river, 143

Medieval tide mill schematic, 56
Mériadec mill, 60
Meteorological tides, 65
Microtidal estuaries, 70
'Mighty Whale", floating power device, 9
Millponds, deterioration of, 45
Mini chemical factory, battery, 17
"Mini-OTEC ", 13
Mini-plants, China, 83
Mixed tide, 66
Moving averages, 73
M/V *Sea Power*, 12

National Wave Energy Research, Development and Administration center, Oregon, USA, 10
Natural barrage, 137
Neap tide, 66
"Nodding duck" (rotating vane), 13
Nuclear plants, 139

Ocean
    alternative energy, 104
    energy
        channel tides, tapping of, 131
        coastal zone alternative energy, 19
        converging wave channels, 9
        ebb- and flood-currents, 107
        extra-low head tidal power, 133
        Gulf Stream, 132
        marine biomass conversion, 14
        OTEC, 1, 13, 22, 23
        renewable, 137
        tidal, 29, 67
    Power/Enersis system, 12
    power, environmental objections to, 16
Oil consumption, 134
Oscillating
    hydroplane principle, 150
    water column (OWC), wave-powered pump, 13
Osmosis principle, 16
OTEC (Ocean Thermal Energy Conversion), 1, 13, 22, 23

Passamaquoddy Bay
    Engineering Investigations Report, 121
    Fisheries Investigations Report (1959), 121
    tidal power plant project, 4, 47, 120, 156
Pen Castel mill, Arzon, Morbihan, 62
Penzinskaya tidal power plant, Russia, 127
Perigean
    spring tides, 67
    tides, 66

Petit Traouiéros mill, Côtes d'Armour, 58
POEMS (Practical Ocean Energy Management Systems), 136
Ponds, 41
Port Royal (Nova Scotia), first (1613) tide mill in North America, 32
Power
  generation, 68
  plant performance, 211
  production, from tides, 104
Powerhouse, on barge, 93
Principal
  lunar ($M_2$) tidal component, 71
  solar diurnal ($P_1$) tidal component, 71
  solar ($S_2$) tidal component, 71
Promising tidal sites, geographical distribution, 144
Proxigean spring tides, 67
Pumped storage, 72

Quoddy
  tides, 122
  undertaking, 119

Raishakou tidal power station, China, 97
Rance Estuary, France, 121
Rance River
  plant, 84
  plant, anatomy of, 109
  plants, 156
  power station
    cross-sectional view, 118
    interior, 117
    site, inside cofferdams, 88
    tidal power plant (Brittany), 2, 30
Rann of Katchchh, 82
Reciprocating turbines, 134
Reconstructed operating tide mill, Southampton, UK, 56
Regional potential, 144
Relict tide mills, density distribution map, 48
Removable
  barrage, 125, 154
  dam, 129
Renewable energies, 137
Rhode Island mill, USA, 42
Rim-type generator, 25
Ruppelmonde mill, West Flanders, 46

St. Banes tide mill, 53
St. Osyth tide mill, Essex, 51
Salford Transverse Oscillator, 148
Salinity gradients, 16
Salt-pan development, 34

San José tidal power plant, Bay of Cadiz, 52, 123, 124
Savonius-type rotor, operating pump, 10, 145
'Sea Clam', 13
Seaflow project, 149, 218
Semidiurnal tide, 66
Severn River
  proposed tidal power plant, 79, 90
  tides, 2
Shunte River, China, 85
Sibulu Strait, Philippines, exploitable power, 145
Single-effect Straflo turbine, 85
Slack period, tidal cycle, 143
Slade's Spice Mill, Chelsea, Massachusetts, USA, 44
Sluiceway caisson, 100
Small tidal power plants, China, 129
SMEC Developments, 131
Smoothing, statistical technique, 73
Solar day, 67
Solsticial tides, 66
Spring tide, 66
Stambridge Mill, England, 42
Statistical techniques
  auto-correlation, 73
  auto-regression, 74
  Fourier analysis, 75
  moving averages, 73
  smoothing, 73
  trend analysis, 73
Stingray project, 150
Storms, risk factor, 151
Straflo
  generating caisson, 99
  turbine, 2, 107, 116
Straits of Dover, 132
Sugar dehydration, 18
Syzygy, 66

Tagus estuary, highest concentration of Atlantic tide mills, 34
Theoretical
  energy, 123
  tidal energy, San José tidal power plant, 123
Tidal
  current
    energy, tapping of, 104
    harnessing of, 131
    power, 66, 67
    resource, 15, 112
    river energy, 112
    turbine, 151, 208

Index

day, 66
energy
    developers, 218
    into power, 103
gates, 132
harmonic components, 71
mill, synchronous starting, 113
movement, potential energy, 67
phase difference, 137
phenomena, 142
power
    centrals, traditional types, 21
    environmental benefits of harnessing, 208
    generating devices, 80
    generation, northwestern Australia, 130
    installations, 211
    major plant sites, 87
    plant, 2
        appropriate site characteristics, 74
        ecological assessments, 157
        environmental concerns, 157
        generating cycles, 134
        Korea, 102
    stations, 108
prism, 71
range, 66, 107
stream, 67
Tide
    generation, 67–70
    mill machinery, 54
    mills
        Aber Wrac'h, 105
        Ancillo, 61
        Bartivas, 59, 60
        Birdham, 50
        Birlot, 36
        Bishopston, 41
        Carew, 49
        Chelsea, 38
        Demi-Ville, 146
        Dunkirk, 146
        East Medina, 49
        Eling, 56, 57
        Ewing, 36
        Grand Traouieros, 35
        Herrera, 41
        Keroilio, 61
        Kisgalobskaia, 154
        Kislaya, 84, 91, 92, 94, 118
        Kislogubsk, 110
        La Rance, 75, 113, 114
        Leqing, 96
        Mériadec, 60
        Passamaquoddy, 3, 44, 122, 156
        Pen Castel, 62
        Penzinskaya, 127
        Petit Traouieros, 62
        Port Royal, 32
        Raishakou, 97
        Ruppelmonde, 46
        St. Banes, 53
        St. Osyth, 51
        San José, 52, 123, 124
        Slade's Spice, 44
        Stambridge, 42
        Trégor, 45
        Uregna, 58
        Van Wyck, 38
        Venera Ria, 63
        Western Europe, map, 64
        Woodbridge, 46
        Zuicksee, 32
    race, 71
    rip, 71
    slack, 143
    types of, 65
        agger, 66
        anomalistic, 66
        current, 86
        diurnal, 66
        equinoctial, 66
        lunar, 71
        lunisolar, 71
        meteorological, 65
        mixed, 66
        perigean, 67
        proxigean, 67
        semi-diurnal, 71
        solsticial, 66
        tropic, 66
Tide-generating forces, 67
Tides, energy generated by, 29
Traditional fuels, 103
Transmission loss, 72
Traou-Meur mill, Côtes d'Amour, Brittany, 46
Trégor, Brittany, mills used for flax, 45
Trend analysis, 73
Tropic tides, 66
Turbine, Kislaya tidal power plant, 95
Turbines
    bulb, 98, 109, 115, 116, 214
    Darrieus, 128, 135, 148, 153, 211
    Davis Hydro, 15
    helic[oid]al, 209
    horizontal axis, 151, 208
    Kaplan, 109
    low-pressure air, 158

Lynmouth, 149
marine current, 209, 218
reciprocating, 134
Straflo, 2, 85, 99, 107, 116
ultra-low head, 79
vertical axis, 20
Wells, 211
Turbodyne Generator, 148

Ultra-low head turbines, 79
Underwater turbines, 128
Uregna tide mill, Zaporito mole, 58
US Electric Power Institute, 10
Utgrunden wind farm, Baltic Sea, 5

Van Wyck mill, New York, USA, 38
Variable-speed tidal mills, 113
Venera Ria tide mill, 63
Verdant Power, 210
Vertical axis tidal power turbines, 20

Wave energy
  converter, 11
Wave energy
  farms, 11
  power, 8, 211
  pre $21^{st}$-century systems, 24
  tapping of, 141
Wells turbine, 211
Wheels, vertical and horizontal, 42
Wind
  farm, 4, 5
  turbine, 5, 151. 209
Woodbridge tide mill, Suffolk, 46
Woodshed Technologies, Australia, 130, 131, 137
World Energy Council, 137

Zuicksee (The Netherlands), tide mill, 32